# International Cryogenics Monograph Series

**Series Editors**

J. G. Weisend II, European Spallation Source, Lund, Sweden
Sangkwon Jeong, Department of Mechanical Engineering, KAIST, Daejeon, Korea (Republic of)

The International Cryogenics Monograph Series was established in the early 1960s to present an opportunity for active researchers in various areas associated with cryogenic engineering to cover their area of expertise by thoroughly covering its past development and its present status. These high level reviews assist young researchers to initiate research programs of their own in these key areas of cryogenic engineering without an extensive search of literature.

More information about this series at http://www.springer.com/series/6086

Thomas D. Bostock · Ralph G. Scurlock

# Low-Loss Storage and Handling of Cryogenic Liquids

## The Application of Cryogenic Fluid Dynamics

Second Edition

Springer

Thomas D. Bostock
Southampton, UK

Ralph G. Scurlock
University of Southampton
Southampton, UK

ISSN 0538-7051 ISSN 2199-3084 (electronic)
International Cryogenics Monograph Series
ISBN 978-3-030-10640-9 ISBN 978-3-030-10641-6 (eBook)
https://doi.org/10.1007/978-3-030-10641-6

Library of Congress Control Number: 2019935828



This Springer imprint is published by the registered company Springer Nature Switzerland AG
The registered company address is: Gewerbestrasse 11, 6330 Cham, Switzerland

*Dedicated to Maureen*

# Preface to the Second Edition

Since the first edition was published in 2006, the breadth of cryogenic applications and the modelling of cryogenic fluid dynamics or CFD have expanded in several directions.

In this second edition, we have extended most chapters to introduce discussions of these new applications and their safety and energy economy. These include advances in the modelling of CFD required in, for example,

(1) The achievement of cooldown and eventual condensation of a working fluid or product with minimum energy and entropy production and minimal energy loss in the design of miniature cryocoolers and condenser/ reboilers.
(2) Gaining a better understanding of large-scale cryogenic liquid mixture properties and their stability in the face of strong convective behaviour in storage and handling.
(3) And understanding that hazards and safety problems in the public domain increase in magnitude as scales of operation of cryogenic systems increase in size.

Southampton, UK
2018

Thomas D. Bostock
Ralph G. Scurlock

# Preface to the First Edition

It is hard to imagine today how difficult it was in 1954 to use cryogenic liquids for low-temperature research. Liquid helium was made in miniature liquefiers incorporated into individual research cryostats. Each cryostat required both liquid air (no liquid nitrogen in 1954), supplied by the British Oxygen Company, and liquid hydrogen which was produced three times a week in the laboratory central liquefier. On liquefaction days, liquid hydrogen was not available until midday and then only by ballot to determine a position and allocation time in the day's queue. If too far down the list, it meant no liquid hydrogen that day, but a high-priority position for the next liquefaction day. Once you had your liquid hydrogen, you could start to make your personal supply of liquid helium.

This operation took several hours, yielding perhaps 100 or 200 ml of precious liquid helium for the research "run". The time might then be late evening, but experimental work went on through the night on the run.

The Clarendon Laboratory, at Oxford University, then built a small central helium liquefier—and research life started to change for those fortunate enough to be able to use "free liquid helium". Not for us however—until the liquid hydrogen cooled cryostat blew up.

A cryostat using "free liquid helium" was to be built as a replacement, but how to design such a replacement was a mystery—there were no textbooks on cryogenics in 1956. Scott's "Cryogenic Engineering" was the first to be published, in 1959 (with 6 reprints by 1967), since when there have been other publications.

The beginnings of our ventures into cryogenics began from that early "free helium cryostat" which was needed urgently for nuclear orientation research on parity non-conservation using low-temperature particle counters.

"Low-Loss Storage and Handling of Cryogenic Liquids: The Application of Cryogenic Fluid Dynamics" is a text which brings together the fundamentals of what we have learned over some 60 years about cryogenic liquids and their sometimes extraordinary behaviour. During this time, we have come to understand how to store and handle cryogenic liquids with growing confidence and efficiency, to the point today where 100 hour and 100 day containment times are the norm, and 1000 days are practicable, within an enormous range of applications. This compares with the 3 hours available for "runs" 60 years ago, a time when the thought of any application of cryogenics seemed totally impracticable.

Our early lectures on storage and handling cryogenic liquids were continually modified by the sustained thread of research work our M.Sc. and Ph.D. students carried out on cryogenic liquids, with the continuous financial support of the Science and Engineering Research Council. Our research findings have been published over the years in some 180 papers scattered throughout various cryogenic journals, conference proceedings and text books.

This text therefore brings together these findings under the general subject heading of cryogenic fluid dynamics and thereby aims to provide a basis for developing low-loss cryogenic systems today, operating at minimum cost. A great deal of the material is Southampton University based as the major, if not only, a source of research findings on cryogenic fluid dynamics. We have therefore used material from M.Sc. and Ph.D. theses of many of our students, as the main sources of information, data and proofs of concept, within the text.

Southampton, UK
2006

Ralph G. Scurlock

# Acknowledgements

My grateful acknowledgements include, in particular, the following Ph.D. research students in chronological order: P. Lynam, A. Mustafa, M. Wray, W. Proctor, J. Boardman, D. Richards, G. Beresford, O. San Roman, A. Tchikou, R. Rebiai, M. Atkinson-Barr, Y. Y. Wu, S. Mirza, M. F. Wu, T. Agbabi, S. Yun, J. Shi and A. Thomas.

We are greatly indebted to the teaching staff and research fellows of the Institute of Cryogenics, University of Southampton, Prof. C. Beduz, Prof. J. P. H. Watson, Dr. K. Kellner, Dr. I. P. Morton, Dr. N. Richardson, Prof. Y. Yang, Dr. M. S. Islam, Dr. M. G. Rao, Dr. D. Utton, Dr. A. C. R. Tavener, Dr. R. Webb and Dr. M. J. Burton.

We are also much indebted to many University Staff who contributed to or collaborated with the development of cryogenic fluid dynamics and its many fascinating applications, thereby broadening each other's cryogenic research and training activities, including, Prof. G. Lilley and Prof. M. Goodyer, Department of Aeronautics and Astronautics (cryogenic wind tunnels, space applications); Prof. P. Hammond, Dr. R. Stoll and Dr. B. Weedy, Department of Electrical Engineering (superconducting cables, transformers, rotating machines and high current leads): Prof. R. Bell, Prof. R. Farrar, Prof. S. Hutton, Dr. D. A. Wigley and Mr. R. J. Bowen, Department of Mechanical Engineering (low-temperature materials and insulations, pipe freezing); Prof. A. Gambling and Prof. H. Kemadjian, Department of Electronics (cold electronics with 0.01% precision using multichannel 16-bit A/D CMOS converters); Prof. J. Fraser, Department of Surgery (cryosurgery techniques and equipment); Prof. G. A. Hills and Dr. A. J. Rest, Department of Chemistry (low-temperature chemistry and FTIR spectroscopy); Dr. R. E. Craine, Department of Mathematics (CFD modelling of convective mixing and rollover); Capt. G. Angas, School of Navigation, Warsash (handling problems with LNG and LPG sea tanker cargoes).

Finally, I am extremely grateful to my wife Maureen for typing and correcting the many drafts of this text and to Phil Cook for preparing all the figures.

Southampton, UK
2018

Thomas D. Bostock
Ralph G. Scurlock

# Contents

# List of Figures

# List of Tables

# Chapter 1
# Introduction

## 1.1 Background, and Redefinition of "Cryogenics" to Include All Temperatures Below 273 K

Some 60 years ago, our working experience began with cryogenic liquids, firstly oxygen, hydrogen and helium 4, then liquid air, nitrogen and helium 3: and more recently, liquid hydrocarbons, methane, ethane, propane, butane and hydrocarbon mixtures, argon and argon mixtures, and neon.

From 1966, a group of us within the Science and Engineering Faculties, University of Southampton, started to teach cryogenics as a post-graduate Masters course. Part of the instruction included demonstrations and experimental studies with a variety of cryogenic liquids. All Masters students carried out a project in industry for the last three months of the course, and this brought us into contact with industrial problems and practices in the handling of cryogenic liquids. Many of these problems were turned into research projects involving Ph.D. research students and/or post-doctoral research fellows … or contract research work within the Cryogenics Industrial Advisory Unit. All teaching, project, contract and research work at Southampton provided a fund of experience on which lecture notes on cryogenics were honed and polished over the years. This monograph is based on some of those lecture notes together with the wealth of experience built up over the years.

A large proportion of the industrial problems we were meeting, extended upwards in temperature from liquid oxygen at 90 K, to liquefied natural gases at 112 K, to liquid ethylene at 169 K, to liquid propane at 233 K and to liquid butane at 272 K. We soon found that the handling and storage behaviour of liquids at the higher temperatures had a significant commonality with the behaviour of liquid oxygen, nitrogen and helium, the latter being the standard cryogenic liquids used in a low temperature physics laboratory [1]. The old definition of cryogenics relating to temperatures below 120 K, as recommended in 1971 after the Paris meeting on Terminology in Cryogenics [2], was changing through common usage. Industry was



T. D. Bostock and R. G. Scurlock, *Low-Loss Storage and Handling of Cryogenic Liquids*, International Cryogenics Monograph Series,
https://doi.org/10.1007/978-3-030-10641-6_1

driving us into redefining cryogenics as relating to all temperatures below ambient—a unifying step towards the development of the subject of "cryogenic fluid dynamics".

We were not alone because in 1983, at the US Cryogenic Engineering Conference, Colorado Springs [3], Haynes and his theoretician colleagues at NBS, now NIST, Boulder, Colorado, USA, recommended the redefinition of cryogenics to include all temperatures below the triple point of water at 273 K. This monograph therefore adopts throughout, this extended 1983 definition of cryogenics, and includes the storage and handling behaviour of all liquids, including hydrocarbon liquids, with NBPs below 273 K.

## 1.2 Early Experiments with Vapour Cooled Baffles and the Empty LOX Pot

Until about 1956, experiments using liquid helium as a coolant, say for adiabatic demagnetisation studies, required at the Clarendon Laboratory, Oxford University, small integral helium liquefiers as part of the working cryostats. The cryostats were all designed and built in the laboratory workshop since there were no cryostat manufacturers in existence at that time. (See, for example, Refs. [4, 5])

The standard practice with all these cryostats was to have a liquid oxygen pot (LIN was not available as a laboratory coolant in the UK until around 1959) in the upper part of the cryostat, perhaps with a liquid hydrogen pot below. This arrangement reduced heat in-flows to the liquid hydrogen and helium sections lower down the cryostat and also provided a first stage heat exchanger for pre-cooling the helium to be liquefied (such as by a Simon bomb liquefier). A great deal of time was therefore spent making liquid helium in situ prior to an experimental run.

Then in 1956, a small central, liquid hydrogen pre-cooled, helium liquefier was built at the Clarendon Laboratory, from which liquid helium could be transferred either directly, or via an intermediate transfer Dewar, into experimental "free" helium cryostats.

The first of these time saving, free helium cryostats "naturally" incorporated a LOX pot built into the head of the cryostat. However, it was first observed by Dr. Horst Meyer (later Professor of Physics, Duke University, NC, USA) in 1956 that the helium evaporation rate from his new "free" helium cryostat fell noticeably, by some 30–50%, when he accidentally allowed the LOX to boil away completely leaving the pot empty. The rate increased sharply to its previous value, however, when he refilled the LOX pot [6].

Meyer's observation suggested that cold helium vapour was cooling the LOX pot below the boiling point of LOX at 90 K and thereby was somehow reducing the heat flow into the liquid helium. Some simple experiments with flat, horizontal, copper baffles in the top of the neck of a free helium test cryostat indicated that the

LOX pot was unnecessary for obtaining a satisfactory helium evaporation rate … but the significance of this simple finding was not appreciated until several years later.

These were exciting times at the Clarendon Laboratory. Early in 1956, two American physicists, Lee and Yang, Columbia University, New York, USA [7] suggested the weak force of beta-decay does not in fact obey parity conservation. The same year, our nuclear orientation cryostat was modified to use photographic film at a temperature of 1.0 K to measure beta particle intensity. By Christmas 1956, our experimental runs had all confirmed the asymmetry of beta-decay from polarised, $^{60}Co$ when cooled to around 0.01 K by adiabatic demagnetization [8]. This positive result, of parity non-conservation, together with those of Wu, Ambler et al., NBS, Washington, USA [9], and Postma et al., Kamerlingh Onnes laboratory, Leiden, Netherlands [10] helped lead to the Nobel Prize in Physics 1957 being awarded to Professors Lee and Yang, Columbia University. As mentioned by Hawking in "A Brief History of Time" [11], this demonstration of parity non-conservation provided a major step towards remodelling the previous theories of particle physics.

Shortly afterwards, in 1957, the photographic cryostat disintegrated during a hydrogen explosion, caused by an inadvertent blockage by a plug of solid hydrogen in the entrance to a liquid hydrogen pot full of liquid. Our first free helium cryostat was a replacement, designed to use beta particle counters in place of photographic film at liquid helium temperatures. It was probably the first cryostat ever to use vapour cooled, copper baffles in place of the conventional LOX pot in the top of the neck. However, this feature was incidental to the progress of nuclear physics at that time in 1957.

Our first serious venture into cryogenics came about at Southampton University in 1965, when doctoral research student, Peter Lynam, carried out some systematic studies on the effect of vapour cooled baffles on liquid helium evaporation rates.

The studies were spectacularly successful, and demonstrated for the first time how the radiation heat flow down the neck of liquid helium Dewar's could be reduced to a negligible amount. This discovery was patented with the help of the National Research Development Corporation, UK [12], and the results were first reported by Lynam, Proctor and Scurlock in 1966 at the meeting of Commission A3 of the IIR [13] in Grenoble, France.

The results indicated that not more than 4 horizontal baffles in the upper neck were adequate to reduce the radiation heat flow very closely to zero, thereby reducing the liquid helium evaporation rate by a factor of 3 or 4.

The use of plastic foam plugs instead of flat baffles was later found to be significantly inferior [14], a fact which has been sadly missed in the design of some commercial cryostats and Dewars, such as cryobiological specimen storage vessels, which use foam plugs in their necks.

Today, vapour cooled radiation baffles are taken for granted and are incorporated as standard practice in all liquid helium and nitrogen systems, while vapour cooled, "suspended deck" insulations are standard in LNG storage tanks with diameters up to 100 m (see Sect. 4.2 for further details).

## 1.3 Discovery of Unstable Evaporation of Liquid Nitrogen

For many years, we came to accept that the evaporation behaviour of cryogenic liquids included some peculiar and unexpected characteristics, without being seriously concerned, although since 1970, we had been teaching cryogenics, including fluid behaviour.

This casual acceptance stopped abruptly in 1975, when we were approached by the Atomic Energy Research Establishment, Harwell, UK, with a problem concerning explosions with 50 l LIN storage vessels. The Harwell committee of enquiry had been unable to discover any mechanical reason for the Dewar failures; the Dewars had been cold and had contained LIN at the time of each event.

It was not at all clear to us what had happened. However, it had been observed, just previously at Southampton, that some of the student experiments with 1–2 l of LIN had been complicated by the wildly irregular evaporation behaviour of LIN. The experiments were using rotameter-type, wet-gas meters yielding integral or averaged measurements of evaporation rate. Sometimes, the meters stopped (or stuck, we believed) for several minutes, and sometimes they whizzed round at way above the average rate, spewing water out of their exits. AERE, Harwell suggested we set up some monitoring tests of their large storage Dewars, which frightened everyone, but we persisted after putting steel screens around the test Dewars.

Since there was no instrument on the market at that time, the first step was to build an electrically recording, instantaneous gas-flow meter. Using the Hastings principle, of a low thermal conductivity tube with a short section of heater and two thermometers on the tube wall, mounted equidistant upstream and downstream of the heater, the electric flow-meter turned out to have a time constant of about 0.5 s. The first readings with evaporating LIN were plagued (by DC noise, or so we thought) by large, random, rapid variations in flow rate with a time constant in the range of 1–10 s. After a month or so, rebuilding and checking the home-made electronics, we came to realise that the rapid variations, by up to 10% of the average flow rate, reflected the normal irregular evaporation of LIN.

Occasionally, much larger flow rate spikes were observed. Tapping or shaking a Dewar also induced large flow rate spikes. These spikes were the first indication of what are now called "vapour explosions", when the rate of vapour production might possibly exceed the limiting, choked flow capability of the narrow necks of the Dewars used by AERE, Harwell. The solution to the problem therefore appeared to be to use wider-necked Dewars.

At that time of the 1975 AERE, Harwell studies, we were raising more questions than answers.

It was clear we did not understand at all the evaporation behaviour of cryogenic liquids in general, or of LIN in particular. A programme of research was therefore set up, with master and doctoral students taking part under the supervision of post-doctoral research assistants.

With financial support from the Science Research Council, this work continued at Southampton for the next 20 years, covering very productively areas such as:

- The boundary layer flow and recirculating mass flow in the cold vapour above cryogenic liquids at scale diameters from 10 mm to 100 m.
- The enhanced heat transfer in stratified cryogenic vapour columns.
- The use of Laser Doppler Anemometry LDA to measure vapour and liquid velocity distributions in cryogenic fluids.
- The use of flow visualisation techniques employing still- and video-photography to compliment LDA velocity measurements.
- The boundary layer and recirculating flows in cryogenic liquids leading to stratification effects.
- The use of video-Schlieren techniques and micro-thermometry to study micro-convection in the liquid surface during evaporation.
- Some 100 "rollover" mixing events during simulation experiments of mixing between 2 liquid layers of different density using LIN/LA and LIN/LOX mixtures.
- Extensive theoretical and computer simulation work to model the dynamical behaviour of cryogenic fluids during evaporation, stratification, rollover, etc.
- Working experience with Helium 3, Helium 4, Hydrogen, Neon, Nitrogen, Air, Argon, Oxygen, Methane, Ethane, Propane, Butane, Carbon dioxide and mixtures such as Liquefied Natural Gases (LNG) and Liquid Petroleum Gases (LPG).

## 1.4 The Contents of This Monograph

This monograph is therefore based on experimental results obtained over the past 40 years or so, and contains theoretical interpretations to enable explanations to be made of the unstable evaporation phenomena associated with cryogenic liquids.

After a detailed discussion of the surface evaporation process, the monograph looks at the various sources of heat flow into a cryogenic system, and considers how these may be absorbed by the cold vapour, rather than by the liquid, in order to minimise boil-off rates.

The monograph then examines the behaviour of cryogenic mixtures and the problems which can arise with mixtures from stratification and spontaneous convective mixing or "rollover".

A chapter on liquid handling then sets out the need to use sub-cooled liquid, thereby avoiding 2-phase flow, for trouble-free transfer.

A chapter on design gathers together all the ideas discussed in this monograph, and relates them to twelve totally different cryogenic systems, on scales ranging from a 10 mm diameter sample holder, to a 100 m diameter LPG tank.

A final chapter on safety introduces extensive working experience, making many points not covered in Safety Manuals.

The monograph provides a unique basis for the design of low-loss Dewars, containers, vessels, and tanks, together with guidance towards the economical and safe handling of cryogenic liquids; all of which is confirmed by the good housekeeping practices developed at the Institute of Cryogenics and placed in the hands of the many students we have trained.

Before going on to describe the evaporation process, and all the subsequent cryogenic fluid dynamical consequences, some definitions of the terminology describing liquid states as used in this monograph are introduced. These definitions may not coincide with those used by steam engineers, so be warned!

## 1.5 Definitions of Single Component Liquid States

### *1.5.1 The 1983 Definition of a Cryogenic Liquid, with Normal Boiling Point Below 273 K*

In this monograph, time after time you will see that if a liquid is at a temperature below ambient, it is subject to heat inflows from its environment and associated natural convection phenomena, which together determine its storage behaviour and handling properties.

There are no other differences in basic physical properties between ambient temperature liquids, like water or gasoline, with normal boiling points above ambient, and cryogenic liquids with normal boiling points below ambient, (apart from the superfluid behaviour of Helium 3 and Helium 4).

Therefore, for the purpose of this monograph, and in line with the NBS/NIST 1983 recommendation, we have adopted the definition of a cryogenic liquid as one with a normal boiling point below the triple point of water at 0.13 °C, or 273 K. The subject of cryogenic fluid dynamics is then developed throughout the book to include the behaviour of all cryogenic liquids with NBPs below 273 K.

The 1983 definition supersedes the previous definition made in 1971 under the leadership of Nicholas Kurti [2] with its division between refrigeration and cryogenic engineering at 120 K. This division arose because temperatures below 120 K can only be achieved using a single refrigerant by employing regenerative cooling of the compressed refrigerant via a heat exchanger or regenerator prior to its expansion to a lower pressure and temperature.

For an extensive discussion of the 1983 definition of cryogenics, see the first six pages in The History and Origins of Cryogenics, published in 1992 [1], which were prepared and agreed with Nicholas Kurti, the author of the old 1971 definition.

Figure 1.1 shows the general equilibrium or saturation P–T curve, together with the various states to be defined below.

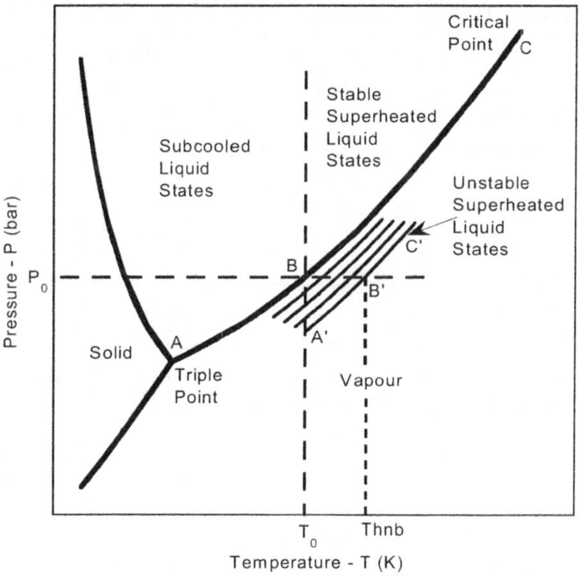


**Fig. 1.1** Typical P–T diagram showing equilibrium curves and thermodynamic states of subcooled and superheated liquid (Thnb = homogeneous nucleate boiling temperature)

### *1.5.2 Boiling Temperature*

Under atmospheric pressure $P_0$, the "boiling temperature" of the liquid $T_0$ is the temperature at which the freely boiling liquid cryogen is in equilibrium with its vapour at pressure $P_0$.

### *1.5.3 Saturation Temperature and Saturation Vapour Pressure*

The boiling temperature $T_0$ may also be called the "saturation temperature", while $P_0$ may be called the "saturation vapour pressure".

### *1.5.4 Normal Boiling Point NBP or Standard Boiling Point SBP at Standard Atmospheric Pressure of 1 Bar*

Since there is a one-to-one relationship between P and T under equilibrium conditions along the saturation P–T curve, if $P_s$ is defined as a standard atmospheric pressure of 1 bar, then the corresponding temperature $T_s$ is also defined as the "normal boiling point" NBP, or the "standard boiling point" SBP (Table 1.1).

**Table 1.1** Basic properties of cryogens at $T_S$ (NBP)

| | MW | $T_M$ (K) | $T_S$ (K) | $T_C$ (K) | $P_C$ (bar) | $\rho_L$ (kg/m$^3$) | $\lambda$ (kJ/kg) | V ratio[a] | $\rho$ gas[a] |
|---|---|---|---|---|---|---|---|---|---|
| Helium 4 | 4 | – | 4.22 | 5.2 | 2.27 | 125 | 20.72 | 738 | 0.14 |
| n-Hydrogen | 2 | 13.8 | 20.28 | 32.94 | 12.84 | 70.8 | 445.4 | 828 | 0.07 |
| Neon | 20.2 | 24.55 | 27.09 | 44.4 | 26.53 | 1205 | 85.7 | 1410 | 0.70 |
| Nitrogen | 28 | 63.15 | 77.31 | 126.3 | 33.99 | 806.8 | 198.9 | 680 | 0.98 |
| Air | 28.9 | – | 78.9 | 132.5 | 37.86 | 875 | 202.6 | 713 | 1.00 |
| Argon | 40 | 83.80 | 87.28 | 150.9 | 49.06 | 1394 | 161.3 | 824 | 1.40 |
| Oxygen | 32 | 54.36 | 90.19 | 154.6 | 50.43 | 1141 | 212.5 | 842 | 1.12 |
| Methane | 16 | 90.69 | 111.7 | 190.5 | 46.0 | 422.4 | 510.3 | 622 | 0.56 |
| Krypton | 83.8 | 115.8 | 119.8 | 209.4 | 54.96 | 2414 | 107.9 | 680 | 2.93 |
| Ethylene | 28 | 104.0 | 169.4 | 282.3 | 50.4 | 568 | 482.6 | 478 | 0.97 |
| Ethane | 30 | 90.35 | 184.6 | 305.3 | 48.71 | 544.1 | 488.5 | 427 | 1.05 |
| Carbon dioxide | 44 | – | 194.7[b] | 304.3 | 73.0 | 1560[b] | 563.0[b] | 823[b] | 1.52 |
| Propane | 44 | 85.45 | 231.1 | 370.0 | 42.1 | 581 | 426 | 312 | 1.52 |
| n-Butane | 58 | 138 | 272.6 | 426 | 36.0 | 601 | 386 | 241 | 2.00 |
| Water | 18 | 273 | 373 | 647 | 217.7 | 1000[c] | 2257 | 1800 | 0.62 |

*Note* [a]V ratio = Volume ratio gas/liquid at 288 K; ρ gas = Gas density relative to air at 288 K
[b]Sublimation values
[c]At 373 K

## *1.5.5 Superheated Liquid*

If the liquid temperature is greater than $T_0$, the liquid is said to be "superheated". Considering this definition more carefully, let us regard Fig. 1.1 as a thermodynamic state diagram and the general liquid state defined by the two co-ordinates P and T, or (P, T). Then we see that two types of superheated state can occur, namely:

1. When the (P, T) state lies above the saturation vapour pressure curve ABC, and to the right hand side of the $T_0$ isotherm. This is a "stable superheated state".
2. When the (P, T) state lies below the saturation curve ABC, along the curves A′B′C′. This is the "unstable superheated state" from which the liquid normally evaporates or changes state.

It will be shown later that the bulk liquid is superheated above $T_0$ by a finite temperature difference, from a fraction of 1 K up to as high as 8 K. This superheated state of bulk liquid in storage is the unstable type, characterised by the curves A′B′C′.

A superheated liquid state is generated when heat is added to the liquid commencing at $T_0$ but without change of phase, and with or without an increase in pressure.

### 1.5.6 Liquid Superheat

"Liquid superheat" is a loose term for the temperature difference ($T - T_0$) above $T_0$, or the enthalpy increase ($H - H_0$) associated with the superheated liquid.

### 1.5.7 Subcooled Liquid

If the liquid state (P, T) lies above the saturation curve ABC and to the left hand side of the $T_0$ isotherm, then the liquid is said to be "subcooled". The term subcooled also applies if the liquid temperature remains at $T_0$ while the pressure is raised above $P_0$.

A subcooled liquid state is generated when:

1. Heat is removed from the liquid, initially at $T_0$, and the degree of subcooling can be specified by the enthalpy removed ($H_0 - H$).
2. The ullage pressure is increased above $P_0$ at constant $T_0$, when there is little change in enthalpy and the degree of subcooling may be specified in terms of the pressure head ($P - P_0$).

### 1.5.8 Wall Superheat

Under heat transfer between solid and fluid, the solid surface must be at a higher temperature than the fluid in contact with, or immediately adjacent to it, to drive the heat into the fluid. The temperature difference to facilitate the heat transfer is called the "wall superheat".

For convective heat transfer to a liquid, the wall superheat is, for example, from 0.01 to 0.1 K in magnitude for LIN, driving low heat fluxes into a liquid boundary layer flow. For convective heat transfer into a vapour, the wall superheat is, for example, of the order of 0.1–1.0 K for cold nitrogen vapour. For nucleate boiling heat transfer, the wall superheat is from around 1 K up to 10 K (or higher for LIN) when the wall heat fluxes can be very large.

### 1.5.9 Boil-off and Boil-off Rate

The terms "boil-off" and "boil-off rate" are strictly applicable only when the liquid is boiling by the heat transfer process of nucleate boiling. In the majority of storage situations, there is only evaporation from the surface of the liquid and there is no boiling. The term "evaporation rate" is then the correct terminology to use.

However, boil-off is in common use to describe all liquid evaporation and is frequently used in this monograph. The boil off rate can be in mass flow, volume flow or % volume flow.

### *1.5.10 Heat Flux and Heat Flow*

The term "heat flux" is the quantitative wording for heat flow per unit area and is usually measured in $kW/m^2$ or $mW/m^2$. Heat flow is the general term measured in kW, W, or mW.

### *1.5.11 Mass Flux and Mass Flow*

The term "mass flux" is the quantitative wording used for mass flow per unit area and is usually measured in $kg/m^2s$. In this monograph, the term is used to describe the "surface evaporative mass flux".

Mass flow is the general term, which is measured in kg/s.

### *1.5.12 Liquid Terminology*

The following shorthand terminology is used throughout the monograph: LHe for liquid helium 4, LH2 for liquid hydrogen, LIN for liquid nitrogen, LA for liquid argon, LOX for liquid oxygen, GOX for gaseous oxygen, LCH4 for liquid methane, LNG for liquid natural gas mixtures of mostly methane with ethane and some higher hydrocarbons. LPG for liquefied petroleum gas mixtures of mostly propane and butane, and CO2 for carbon dioxide.

## References

1. Scurlock, R.G.: History and origins of cryogenics. Oxford University Press (1992)
2. Kurti, N.: Low temperature terminology. Proceedings of the 13th International Congress Reference, Washington D.C. (1971); also Cryogenics **10**, 183 (1970)
3. Haynes, W.M., Kidnay, A.J., Olien, N.A., Hiza, M.J.: States of thermophysical properties data for pure fluids and mixtures of cryogenic interest. Advances in Cryogenic Engineering **29**, 919 (1983)
4. White, G.K.: Experimental Techniques in Low Temperature Physics, 4th edn. Oxford University Press (2002)
5. Hoare, F.E., Jackson, L.C., Kurti, N.: Experimental Cryophysics. Butterworths (1961)
6. Meyer, H.: Private communication (1956)

7. Lee, T.D., Yang, C.N.: Physical Review **104**, 254 (1956)
8. Grace, M.A., Johnson, C.E., Scurlock, R.G., Sowter, C.V.: A demonstration of parity non-conservation in β-decay. The Philosophical Magazine **2**, 1050 (1957)
9. Wu, C.S., Ambler, E., Hayward, R.W., Hoppes, D.D., Hudson, R.P.: Parity non-conservation in β-decay of polarised $Co^{60}$ nuclei. Phys. Rev. **105**, 1413 (1957)
10. Postma, H., Huiskamp, W.J., Miedema, A.R., Steenland, M.J., Tolhoek, H.A., Gorter, C.J.: Physica **23**, 259 (1957)
11. Hawking, S.W.: A Brief History of Time, pp. 77. Bantam Press (1988)
12. Vapour cooled baffles for reducing heat flows into cryogenic liquids. UK Patent, 1965
13. Lynam, P., Proctor, W., Scurlock, R.G.: Reduction of the evaporation rate of liquid helium in wide-necked Dewars. Bulletin of IIR, Commission 1, Grenoble, Annexe 1965-2, 351 (1965)
14. Lynam, P., Mustafa, A.M., Proctor, W., Scurlock, R.G.: Reduction of the heat flux into liquid helium in wide-necked metal Dewars. Cryogenics **9**, 242 (1969)

# Chapter 2
# Evaporation of Cryogenic Liquids

## 2.1 Introduction

This chapter describes how there are three modes of heat transfer within the fluid mechanisms which lead to evaporation of the liquid. These are:

1. Nucleate boiling from wall to bulk liquid, at high heat fluxes,
2. Convective heat transfer from wall to bulk liquid with no boiling, at the low heat fluxes characteristic of normal storage,
3. Conductive and micro-convective heat transfer in the liquid/vapour interfacial region leading to surface evaporation.

The chapter then continues with a discussion on how, under normal storage conditions when surface evaporation is dominant, a superheated state of the liquid is a necessary occurrence.

At the same time as steady-state surface evaporation takes place, unstable evaporation phenomena are common occurrences which need to be understood, controlled and accepted. This chapter will conclude with outlines of **four** of these phenomena, namely:

- normal, irregular surface evaporation,
- vapour explosions via transient high surface evaporation, usually from single component liquids,
- rollover, via continuous high surface evaporation, as a result of spontaneous mixing following stratification into two or more layers,
- boiling throughout the bulk liquid as quasi-homogeneous nucleate (QHN) boiling.



T. D. Bostock and R. G. Scurlock, *Low-Loss Storage and Handling of Cryogenic Liquids*, International Cryogenics Monograph Series,
https://doi.org/10.1007/978-3-030-10641-6_2

## 2.2 Nucleate Boiling from Wall to Bulk Liquid

### *2.2.1 Heterogeneous Nucleate Pool-Boiling*

The mechanism of heterogeneous nucleate pool-boiling on a submerged heated wall is well documented [1]. Because vapour bubbles have an increased internal vapour pressure (proportional to the ratio of surface tension/bubble diameter), they have an increased saturation temperature which must be exceeded for the bubble to grow. The increase in wall temperature needed to create vapour bubbles in the first place can be reduced by nucleation centres in, or on, the surface of the wall.

Heterogeneous nucleate boiling on plain heated surfaces can be significantly enhanced today by treating the surfaces so as to create a wide variety and a high density of nucleation sites, for example, by the application of porous coatings [2].

For LIN, heterogeneous nucleate boiling on plain surfaces occurs between heat fluxes of a minimum of about 10 $kW/m^2$, rising to a maximum or critical heat flux of about 500 $kW/m^2$, with wall superheats from about 0.5 K to a maximum of about 20.0 K, respectively, the actual values depending on the particular surface and its immediately previous, thermal history.

For LHe, steady state, heterogeneous nucleate pool boiling on plain surfaces occurs between a minimum of about 10 $W/m^2$ rising to a maximum or critical heat flux of around 10 $kW/m^2$, with wall superheats from about 0.05 K to a maximum of about 0.8 K [3].

For extended vertical surfaces, like those in reboiler-condensers or the walls of large tanks, the integrated heat transfer into the liquid over a large vertical distance depends significantly on whether the heat transfer takes place at constant wall temperature (the reboiler-condenser condition, which is very difficult to simulate experimentally) or constant wall heat-flux (the usual experimental simulation), and whether the bulk liquid temperature is constant with depth, or not.

For LIN, $T_s$ rises by 1 K at a liquid depth of 14 m. nucleate boiling will therefore tend to be suppressed at greater liquid depths, particularly under the constant wall temperature condition in a reboiler-condenser, rendering the lower part ineffective for heat transfer purposes.

This suppression of nucleate boiling with liquid depth is not observed with constant wall heat flux, experimental boiling heat transfer rigs. A sophisticated computer controlled, variable heat flux with height, constant wall temperature, experimental rig will, however, demonstrate quite dramatically the boiling heat transfer ineffectiveness of the lower part of the rig in an isothermal pool of LIN [4, 5].

For heterogeneous nucleate boiling in a liquid at its saturation temperature $T_0$, streams of vapour bubbles rise to the surface and break through to become the boil-off vapour mass flow.

If the liquid is subcooled with respect to $T_0$, the rising vapour bubbles may collapse due to recondensation of their vapour content, so that there is no boil-off vapour. Instead, all the latent heat of condensation of the bubbles goes into heating the bulk liquid.

### 2.2.2 High Efficiency Heterogeneous Nucleate Boiling Heat Transfer Using Falling Liquid Films

In an air separation unit (ASU) reboiler/condenser, nitrogen vapour under pressure condenses against boiling oxygen under a $\Delta T$ of several °C. This $\Delta T$ is determined largely by the nucleate boiling heat transfer coefficient. It can be significantly reduced by using enhanced porous surfaces, but is limited by the increased subcooling introduced by the liquid head of the LOX in the reboiler.

The constant wall temperature heat transfer can be enhanced, together with zero subcooling, by replacing the LOX bath with a pumped falling film of boiling liquid. The falling film technique has been found to reduce $\Delta T$ and increase the heat transfer coefficient by up to 50 fold. Although a pumped falling film is a mechanical complication, requiring a filter to remove impurities in the LOX, the net energy consumption is significantly reduced.

### 2.2.3 Homogeneous Nucleate Boiling

In the absence of nucleation sites, a liquid can be heated by heat fluxes normally associated with heterogeneous nucleate boiling to a temperature $T_w$ well in excess of $T_0$ without boiling. With continued heating of the liquid, the excess superheat $(T_w - T_0)$ can become much larger than that for heterogeneous nucleate boiling and large enough for homogeneous nucleate boiling to take place. Vapour bubbles then form spontaneously throughout the liquid and grow very rapidly, almost explosively. Large volumes of vapour mixed with liquid are produced, and a mixture of vapour and liquid may be carried out through the vents and the safety valves. The associated pressure rise due to the vapour generation may damage the liquid enclosure vessel.

For LIN, the homogeneous nucleation boiling temperature $T_{hnb}$ corresponds to an excess superheat of about 40 K for homogeneous nucleate boiling to take place in the violent fashion described above.

### 2.2.4 Quasi-homogeneous Nucleate Boiling, QHN Boiling

In many situations, similar violent boiling can take place with much smaller unstable superheated states than those required for homogeneous nucleate boiling, e.g. for LIN with superheats of only 1 K or 2 K. These events are termed quasi-homogeneous nucleate (QHN) boiling events, and take place in the absence of solid surfaces with nucleation sites [6].

Bubble nucleation during QHN boiling is believed to take place on either

(a) particulates suspended in the liquid, such as $CO_2$ snow or $H_2O$ ice,
(b) vapour trails created by the passage of cosmic radiation particles through the liquid, or
(c) focussed sound waves, or pressure oscillations from pulsed acoustic sources, such as pumps, valves, and instrumentation.

In all cases, when QHN boiling takes place, there is a large volume of vapour generated throughout the liquid: and a rapid rise in the liquid/vapour interface, with the possibility of liquid being carried out of the vapour vent lines.

In the case of small vessels like liquid helium and nitrogen laboratory Dewars, this flow of two-phase mixture of liquid and vapour out of the vent is similar to that of geysering events and may empty the vessel of liquid.

If the rate of vapour generation exceeds the choked flow discharge capacity of the vent lines and safety valves together, the pressure will rise rapidly until the enclosure fails mechanically.

There is a major problem in detecting the build-up of unstable superheated states in the liquid, or assessing in advance the magnitude of a QHN boiling event, if the liquid temperature is not adequately monitored. This is because the vapour pressure over the liquid is no guide to the degree of superheating of the liquid (see Fig. 1.1). A fall in evaporation rate below the "normal" storage value could be a possible warning guide.

As an example of a QHN boiling event, a pressurising LIN cryostat experiment, at Southampton University for Masters students, accidentally demonstrated this nasty problem very clearly on several occasions. In this experiment, about 2 l of LIN were heated in a 4 l, closed volume vacuum insulated vessel, via a 30 W wall heater consisting of wire wound around the outside of the pressure vessel, so as to measure the $(P_0,T_0)$ saturation curve. The heater wire was not in direct contact with the LIN and presumably its heating effect did not generate any heterogeneous nucleate boiling. Normally, both vapour pressure and liquid temperature increased smoothly with time yielding the standard saturation curve.

However, on several occasions, energising the wall heater caused the LIN to superheat into the unstable superheated state, without any increase in pressure. If the student did not appreciate quickly the significance of his observations of rising temperature but with no corresponding rise in pressure, then after some 30 min, the cryostat would be in a dangerous condition.

Recovery was achieved by turning off the wall heater, evacuating the laboratory, opening the vent valve, and running. The action of opening the vent valve, however slowly, generated a QHN boiling event; the vapour pressure gauge was observed to go off scale above 30 bars in agreement with the following simple calculation.

A heat input of 30 W into 2 l of LIN for 30 min will increase the enthalpy of the system by 54 kJ. This corresponds to a liquid superheat of about 13 K in 2 l to a liquid temperature of 90 K. This heat is absorbed by the flash evaporation of some 200 l of vapour. If the evaporation is by QHN boiling in a very small time, of, say, less than a second, then it will result in a pressure of 30–100 bars being generated in the closed volume of 2 l before the safety valves can open!

## 2.3 Convective Heat Transfer Without Evaporation at the Point of Heat Influx

In most, if not all, storage situations, the heat flow through the insulation into the liquid is a much more gentle process with a heat flux of typically less than 100 W/$m^2$ for LIN. This level of heat flux is some 2 orders of magnitude less than the minimum required for heterogeneous nucleate boiling, and can only be released from the liquid by surface evaporation, with no boiling at all.

To get to the surface, the heat in-flow is first absorbed by a process of natural convective heat transfer creating an upward flow of less dense superheated liquid: there is no boiling and also no evaporation at the point where the heat is absorbed. At a vertical wall, the flow of superheated liquid assumes the form of a boundary layer immediately adjacent to the wall, in a layer about 1–5 mm thick. Heat transfer from the heated wall to the liquid by such a boundary layer flow is very effective and is well documented in many texts on heat transfer [7, 8].

For a cryogenic liquid, the boundary layer flow at the wall absorbs all the heat flow entering the liquid. Furthermore, for a container which has a liquid depth/diameter ratio greater than about 0.5, the heat flow through the base is absorbed convectively by a boundary layer flow across the base which is continuous with the vertical wall boundary layer flow via a boundary layer suction process.

For depth/diameter ratios less than 0.5, such as in large cylindrical LPG and LNG tanks, the boundary layer flow across the base may be broken by the creation of vertical "thermals" spaced horizontally at intervals approximating to the liquid depth according to Rayleigh's instability criteria for natural convection. In all cases, the heat in-flow is carried by boundary layer flows, and thermals, to the liquid surface.

## 2.4 Surface Evaporation

After the superheated boundary layer flows and thermals reach the surface, some, or all, of the excess heat is absorbed by the process of surface evaporation. In fact, the wall boundary flow turns through 90 °C and moves radially inwards just below the liquid/vapour interface. During this inward radial motion, surface evaporation takes place as described in Sects. 2.4.1–2.4.5 (see also Fig. 2.1).

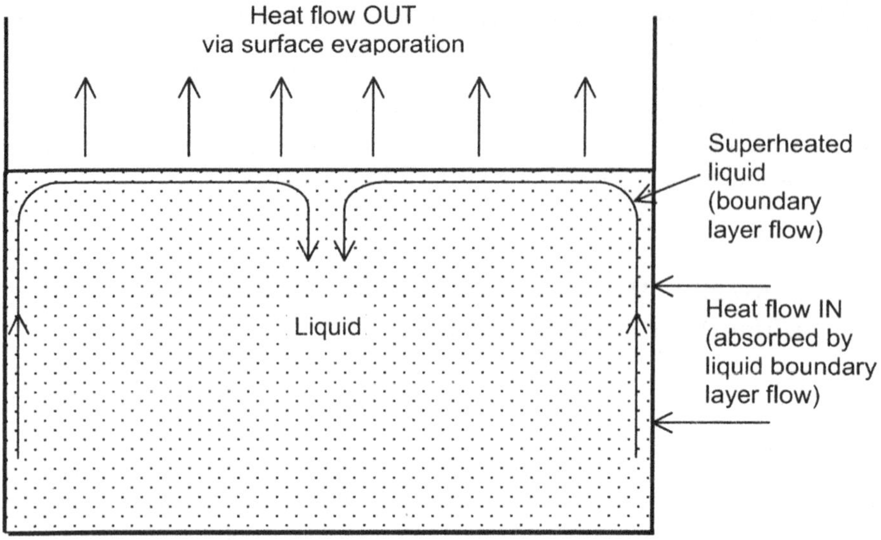


**Fig. 2.1** Surface evaporation whereby heat flow into liquid at the walls is rejected remotely via vapour from the surface

When the inward radial motion reaches the centre, it turns through another 90 °C to become a strong downward jet carrying the excess heat, not released by surface evaporation, into the core of the liquid where secondary convection produces mixing and superheating.

The highest fluid temperature is undoubtedly at the wall–liquid–vapour interface, where the evaporative mass flux will be larger than at the centre of the liquid pool. Surface evaporation is much more complicated and sensitive than this simple description. Let us develop the picture as it was discovered at Southampton.

### 2.4.1 Surface Evaporation Mass Flux and Bulk Superheat

To gain an understanding of the relationship between evaporation mass flux and bulk superheat, microthermometer studies were first made on LIN, LOX, LA, $LCH_4$ and LNG [9–12].

The boil-off vessel was an 80 mm inner diameter, double walled, vacuum insulated Dewar surrounded by a second liquid bath 120 mm inner diameter. The boil-off from the inner vessel could be varied via a uniform heat-flux electrical heater mounted in the vacuum space around the inner wall. The micro-thermometers consisted of 25 μm diameter copper/constantan thermocouple junctions mounted horizontally in differential or absolute configurations (see Fig. 2.2).

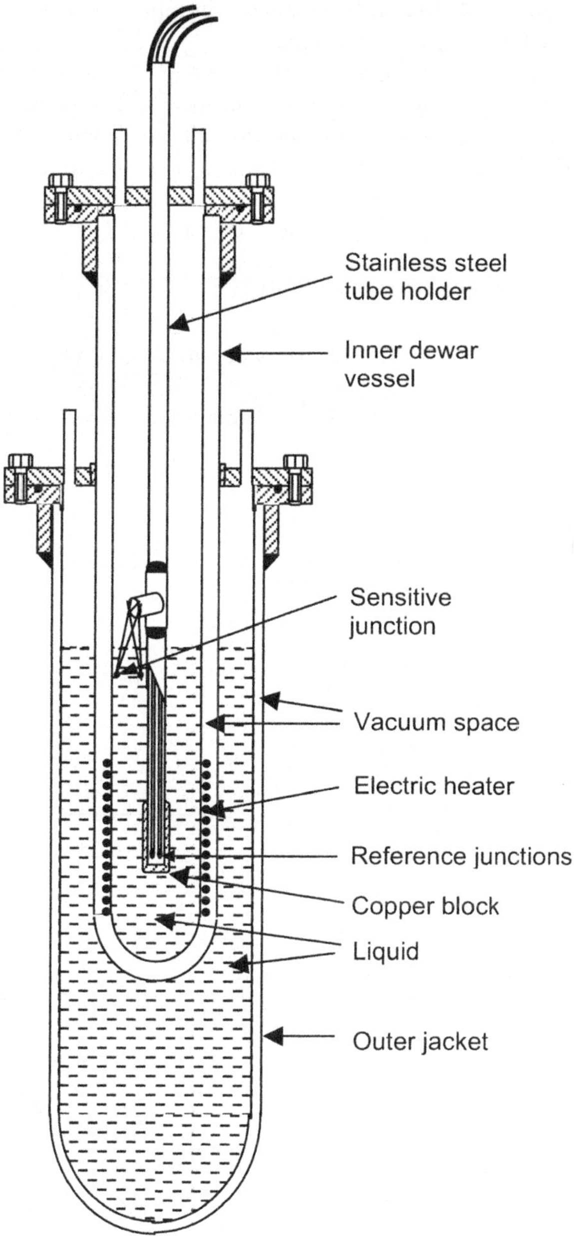


**Fig. 2.2** Experimental rig for evaporation studies

For the differential configuration, the two junctions were separated vertically by a distance of 100 mm, so that temperatures in the region of the surface could be measured relative to that of the bulk liquid. For the absolute configuration, the reference junction was the ice-point while a single calibration point was made against a Platinum thermometer in rapidly boiling LIN. The pool depth was kept between 200 and 250 mm.

The results are summarised in Fig. 2.3 as a log-log plot of average mass-flux through the surface in g/m$^2$ s against bulk superheat $\Delta T$ ($T_b - T_0$). With the uniform heater used, it proved possible to generate superheats of 3.0 K in LIN and LNG, 4.0 K in LOX, and 4.9 K in LA, before any nucleation occurred.

$$\text{For LIN} \quad m = 3.2\,\Delta T^{1.80}\ \text{g/m}^2\,\text{s} \tag{2.1}$$

$$\text{For LOX} \quad m = 3.4\,\Delta T^{1.80}\ \text{g/m}^2\,\text{s} \tag{2.2}$$

$$\text{For LA} \quad m = 3.85\,\Delta T^{1.60}\ \text{g/m}^2\,\text{s} \tag{2.3}$$

$$\text{For LNG (North Sea)} \quad m = 1.60\,\Delta T^{1.33}\text{g/m}^2\,\text{s} \tag{2.4}$$

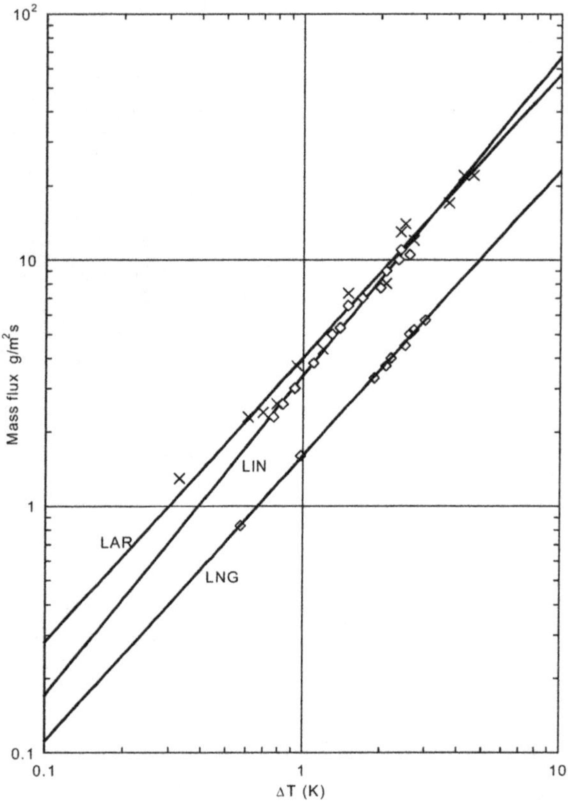


**Fig. 2.3** Evaporation mass flux versus bulk superheat $\Delta T$ for LIN, LAR and LNG (North Sea)

### 2.4.2 Impedances to Surface Evaporation: The 3 Regions in the Surface Sublayer

The impedance mechanisms are revealed by the vertical temperature profiles and their variation with time within a few mm of the liquid-vapour interface, which we shall call the surface sub-layer.

To produce these time dependent profiles, the liquid surface was allowed to fall and pass the rigidly fixed microthermometer system as the liquid evaporated. An example of the vertical profiles in LIN is shown in Fig. 2.4 for a mass flux of 12 g/m$^2$ s, equivalent to a heat flux of about 2.4 kW/m$^2$ through the surface. Similar vertical profiles were observed in surface sublayers of all the liquids, namely LIN, $LCH_4$, LOX, LA and LNG (North Sea).

As can be seen in Fig. 2.4, the surface is cooler than the bulk liquid while the actual vertical temperature profile has a complex structure in the sub-layer down to a depth of about 5 mm, below which the temperature is uniform at $T_b$. These observations are not unique to cryogenic liquids. Oceanographers have a problem with determining sea-water bulk temperatures from infra-red scanning radiometers on satellites. The satellite sensors measure the sea-surface skin temperature characteristic of the top layer less than 100 μm thick, which can be up to 1 K cooler than the bulk on a calm night in the absence of solar heating [13].

Returning to Fig. 2.4. the profile within the sub-layer is dominated by the superposition of many temperature pulses, or spikes, which become observable with a fast response recorder. Expanding the time scale, these spikes can be seen to show an irregular time dependence and include short cold temperature pulses (but

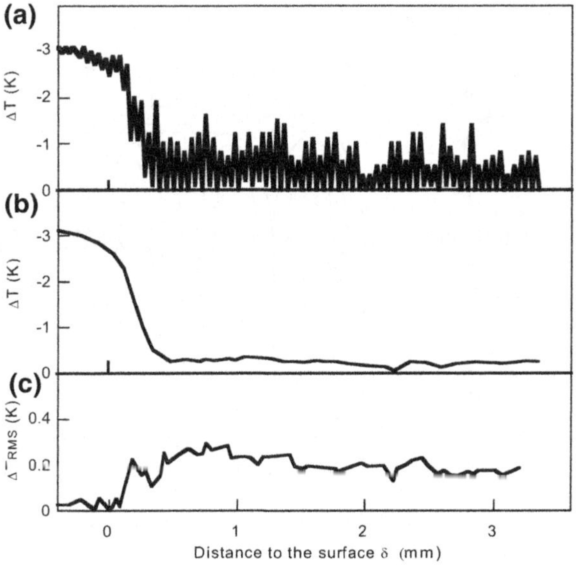


**Fig. 2.4** Microthermometer studies of surface morphology during evaporation, **a** Local temperature variation $\Delta T = T - T_b$ with depth δ below liquid surface. **b** Smoothed variation with δ. **c** RMS variation of fluctuations with δ

never colder than $T_O$) and longer hot temperature pulses (but never hotter than $T_b$). However, the time independent part of the profiles can be separated into 3 regions (working down from the surface) with different temperature gradients, as shown in Fig. 2.5 namely:

(1) A molecular evaporation region at the surface which is probably no more than 1–2 μm in thickness, but appearing to extend to 50–100 μm in practice, as the capillary film remains attached to the thermocouple junction by surface tension forces when the liquid surface falls below it.
(2) A thermal conduction region enhanced by some convection, about 400 μm thick, with an extraordinarily high temperature gradient.
(3) An intermittent convection region, about 5000 μm or 5 mm thick, with a small temperature gradient, which contains the bulk of the observed thermal spikes. The thermal spikes also extend into region 2, but rapidly reduce in intensity as the surface is approached.

**In region 1**, at the vapour/liquid interface of 1–2 μm thickness, molecular evaporation takes place according to the relation [14]:

$$M = 90\,\alpha(T_s - T_O)\,\text{kg/m}^2\text{s} \tag{2.5}$$

This is equivalent to a heat flux $Q_{ev}$ given by:

$$Q_{ev} = m\lambda = 90\lambda\alpha(T_s - T_O)\ \ \text{kW/m}^2 \tag{2.6}$$

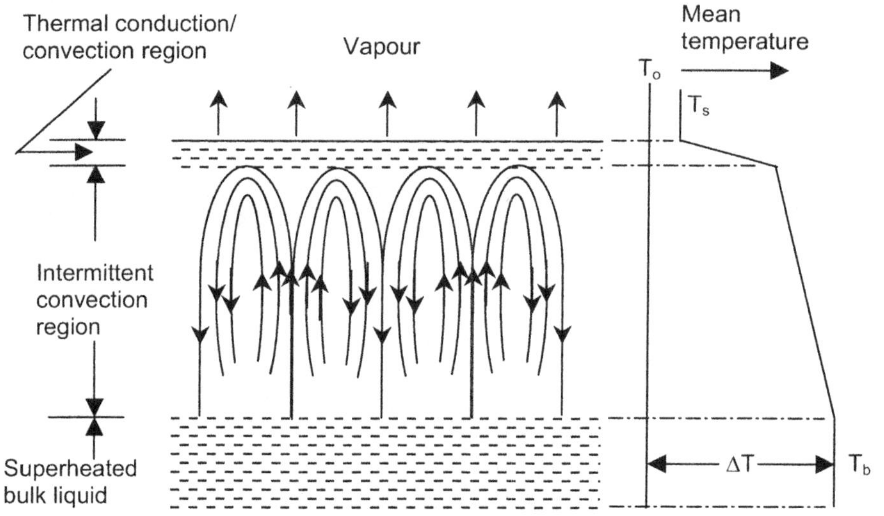


**Fig. 2.5** Morphology and temperature profile across three regions of surface sub-layer of liquid nitrogen

where $T_s$ is the temperature of the vapour/liquid interface, $T_0$ is the saturation temperature, $\lambda$ is the latent heat of evaporation and $\alpha$ is the molecular evaporation coefficient. The reciprocal of $\alpha$ can be regarded as an evaporation impedance factor, the magnitude of which is very sensitive to the concentration of impurities in the surface.

Figure 2.5 shows the temperature profile across the 3 layers between the vapour side of the interface at $T_0$ and the bulk liquid at $T_b$ for LIN at m = 12 g/m$^2$ s where the bulk superheat ($\Delta T = T_b - T_0$) was 3.0 K.

The estimated value of $(T_s - T_0)$ is 0.15 $\pm$ 0.10 K, close to the lower limit of measurement. Inserting values into Eq. (2.5), it can be seen that this measurement leads to an estimated minimum value of the evaporation coefficient $\alpha$ of about $10^{-3}$ for liquid nitrogen.

For uncontaminated water, the observed surface evaporation coefficient is around 0.04. However, it is well known that, in the presence of a monolayer of molecular impurity with a low vapour pressure, the surface evaporation coefficient can fall to the order of $10^{-6}$. This will effectively stop the evaporation of water droplets in the atmosphere, and lead to the occurrence of persistent smogs.

Likewise, if impurities collect at the surface of a cryogenic liquid, the evaporation coefficient can be expected to fall by several orders of magnitude to the order of $10^{-6}$. In other words, the surface evaporation impedance rises by several orders of magnitude and stops evaporation through the contaminated surface of the cryogenic liquid.

**In region 2**, the major temperature drop occurs across a liquid layer of approximately 400 μm thickness with an extraordinarily high temperature gradient of 5000–7500 K/m.

Heat flow across this layer is by a mixture of thermal conduction and highly damped intermittent convection as shown by the diminishing size of the thermal spikes on approaching the surface.

The heat flux for liquid nitrogen is given by:

$$Q_{ev} = m\lambda = k_{eff}(dT/dz) = k_{eff}(T_c - T_s)/\delta(2) \quad (2.7)$$

where $\delta$ (2) is the thickness of region 2. For the experimental example with LIN (see also Table 2.1), the value of $k_{eff}$ is 0.37 W/m K compared with the static conductivity of 0.133 W/m K. Thus we see that the conductance across the region 2 is enhanced by the intermittent convection by a factor of about 2.8.

Table 2.1 shows how this degree of enhanced conductance across region 2 in the evaporative sub-layer applies to the other cryogenic liquids studied, varying from 2.0 for LOX and LA, down to 1.5 for $CH_4$ and LNG.

The heat flow across region 2 for the LIN example is then:

$$m\lambda = 0.8(T_c - T_s)\ kW/m^2 \quad (2.8)$$

where $T_c$ is the temperature at the bottom of this layer of mixed conduction and intermittent convection.

**Table 2.1** Experimental data on effective thermal conductivity in region 2 of surface sub-layer during surface evaporation

| | Surface mass flux ($g/m^2$ s) | ΔT (K) | Temperature gradient ($\times 10^3$ K/m) | $K_{eff}$ (mW/m K) | $K_{actual}$ (mW/m K) | $K_{eff}/K_{actual}$ |
|---|---|---|---|---|---|---|
| LIN | 13.9 | 3.2 | 7.5 | 370 | 133 | 2.8 |
| $LCH_4$ | 3.0 | 2.5 | 5.0 | 305 | 189 | 1.6 |
| LOX | 9.8 | 2.4 | 6.7 | 305 | 152 | 2.0 |
| LA | 14.6 | 3.2 | 10.0 | 236 | 128 | 1.8 |
| LNG | 2.7 | 2.1 | 5.0 | 275 | 189 | 1.5 |

**In region 3**, the depth interval 0.2–5 mm, the temperature difference across this relatively thick layer is only 0.25 K corresponding to a mean temperature gradient of about 50 K/m, much smaller than in the region 2.

The major feature is the temperature spikes which have a time constant of about 0.3 s and a mean amplitude of approximately +0.5 and −0.5 K respectively within the temperature interval $T_b$ to $T_s$. For liquid depths $\delta < 0.4$ mm (i.e. <400 μm) within region 2, the temperature spikes decrease in amplitude to zero as the surface is approached at $\delta = 0$. For $\delta > 5$ mm, at depths below the surface sub-layer, the temperature spikes appear to cease.

It is deduced from these profiles, and subsequent video Schlieren observations (see below in Sect. 2.4.4.), that these temperature spikes arise from convection cells carrying heat across the 5 mm thickness of region 3 under the small temperature gradient observed. The cells appear to consist of narrow falling plumes of cold liquid and wider rising plumes of hot liquid co-ordinated into convecting sheets or streamers moving parallel to the surface past the stationary microthermometers. The streamers have a vertical dimension of 4–5 mm, i.e. the thickness of region 3, and are driven primarily by the cooling and sinking, through increase in density, of "spent" elements of liquid, from the mixed conduction/convection region 2, in the temperature gradient of about 50 K/m.

At the same time, through mass continuity, the streamers incorporate elements of superheated liquid from immediately below region 3 which rise and are transferred to region 2 where they give up their superheat.

This lower thick region, region 3, which we call the intermittent convection layer, therefore contains the mechanism whereby superheated liquid from the inward radial flow of the bulk convection loop (driven by the wall boundary layer flow) is carried into the surface layer by a collective system of rising and falling plumes or convective streamers. This mechanism is not unique to cryogenic liquids and was first identified by Rayleigh from his studies on the evaporation of water. He was seeking to understand how superheated bulk liquid can flow convectively upwards to the colder surface, against the negative temperature gradients and positive density gradients.

For the LIN example, the effective linear heat transfer equation for this intermittent convection layer is given by:

$$Q_{ev} = k_{eff}(T_b - T_c)/\delta(3) \tag{2.9}$$

where $k_{eff}$ = 48 W/m K, $\delta$ (3) is the thickness of region 3, equal to 5 mm, and $T_b$ is the bulk temperature of the superheated liquid. Hence, for the example,

$$Q_{ev} = m\lambda = 9.6(T_b - T_c)\ \text{kW/m}^2 \tag{2.10}$$

It can be seen that the effective thermal conductance of this layer due to the convection process, is about 360 times the thermal conductivity of the static liquid. Combining Eqs. (2.6), (2.8), and (2.10), we obtain the overall heat flux equation for the surface evaporation of liquid nitrogen in the example, where $\lambda$ = 199.0 kJ/kg and the evaporation mass flux, m is in kg/m$^2$:

$$\begin{aligned}(T_b - T_0) &= (T_s - T_0) + (T_c - T_s) + (T_b - T_c)\\ &= m\left[\underset{(1)}{1/90\,\alpha} + \underset{(2)}{250} + \underset{(3)}{0.02}\right]\end{aligned} \tag{2.11}$$

For the liquid methane example in Table 2.1, the surface evaporation impedance equation, with $\lambda$ = 512 kJ/kg and effective thermal conductivity (in region 2) $k_{eff}$ = 0.30 W/m K compared with the static value of 0.19 W/m K, becomes:

$$(T_b - T_0) = m\left[\underset{(1)}{1/35\,\alpha} + \underset{(2)}{545} + \underset{(3)}{0.1}\right] \tag{2.12}$$

The impedance terms represent respectively the contributions from:

(1) molecular evaporation including impurity effects in region 1,
(2) the mixed thermal conduction/convection layer, region 2,
(3) the intermittent convection layer, region 3.

### 2.4.3 General Conclusions from Experimental Studies of Separate Impedance Terms

It is perhaps unfortunate that the analyses of surface evaporation have continued long after the experimental results were completed. It is therefore not possible to go back and change the mass fluxes to get a more complete picture.

However, the temperature profiles obtained at the time enable the following conclusions to be made:

(a) The intermittent convection in region 3 is a common feature of all the cryogenic liquids studied, including LIN, LA, LOX, $LCH_4$ and LNG (North Sea natural gas).

(b) The steep temperature gradients associated with the mixed thermal conduction/convection region 2 are approximately the same (to within a factor of 2) for the various liquids as shown in Table 2.1.
(c) The observed thickness of the mixed conduction/convection layer is about 0.4 mm, or 400 μm, for every liquid.
(d) The important conclusion is that the major impedance to evaporation is provided by the mixed conduction/convection region. Convection via intermittent plumes in region 3 is much more efficient, with an effective thermal conductance about 360 times greater than that in region 2.
(e) Even if the evaporation coefficient $\alpha$ is as small as $10^{-3}$, the impedance term (1) is still over 20 times smaller than term (2). In other words, the molecular evaporation process is not the rate limiting mechanism during normal surface evaporation.
(f) However, this equilibrium process, whereby bulk superheated liquid is separated from the surface by a thin sub-layer in which the heat flux to the surface is almost totally controlled by a mixed conduction/convection layer only 400 μm in thickness, is sensitive to being dramatically disturbed in two totally different ways, firstly, by agitation of the bulk liquid and, secondly, by impurities in the surface.
(g) Agitation of the bulk liquid can lead to liquid motion in the surface sub-layer circumventing or bypassing the mixed conduction/convection region. In this case the rate limiting impedance falls to that of the molecular evaporation, and the overall evaporation impedance **reduces** from $\sim 250$ to $\sim 11$ resulting in an evaporation mass flux increase of 23 fold for LIN. For $LCH_4$, the overall evaporation impedance reduces from $\sim 545$ to $\sim 28$ with a mass flux increase of 19 fold. This is believed to be what happens to the boil-off during a vapour explosion over a short period; or during a rollover if the agitation of the surface sub-layer due to penetrative convection mixing is maintained over a longer period.
(h) Impurities collecting on the surface, either by condensation from the ullage vapour, or from solute remaining behind during evaporation of the solvent liquid, can cause $\alpha$ to fall to the order of $10^{-6}$. In which case, the rate limiting impedance is determined by the reduction in $\alpha$ and **rises** by 30 fold in the case of LIN, and by 40 fold in the case of $LCH_4$, causing the evaporation to almost stop. Possible candidate impurities include water and carbon dioxide.

### *2.4.4 Schlieren Studies of the Surface Interface*

The cryogenic Schlieren system shown in Fig. 2.6 was used in conjunction with a TV camera and video recorder to observe the convective motion within the surface sub-layer directly. A parallel beam of white light from a point source is directed onto a stainless steel mirror submerged 10–30 mm deep in the cryogenic liquid, i.e.

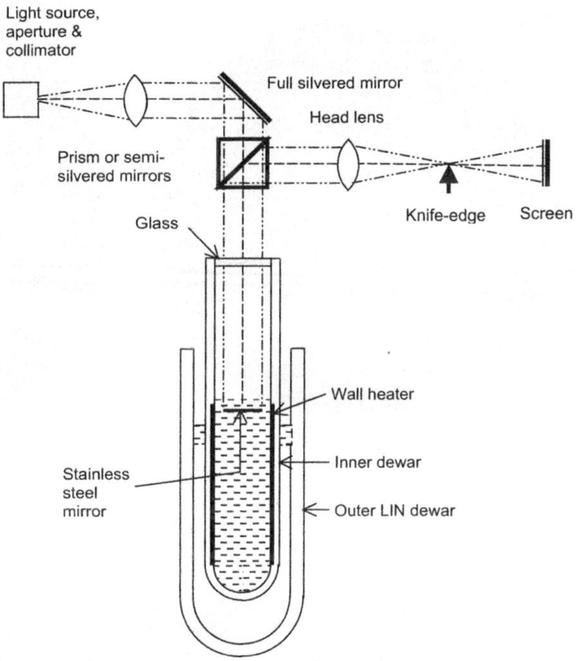


**Fig. 2.6** Experimental setup for Schlieren observation

sufficiently deep to be clear of both the surface evaporation sub-layer and the radial inflow across the surface of superheated liquid from the wall heated boundary flow. The reflected beam of white light passes out and is focussed on to a viewing screen or camera, after passing a knife-edge to remove the undeviated beam.

Bearing in mind that the vertical axis Schlieren optics detect horizontal gradients in refractive index, which relate directly to changes in density produced by horizontal gradients in temperature, local convection patterns are revealed in extraordinary detail.

Figure 2.7 illustrates the instantaneous pattern of radial convection lines or streamers which are revealed. These lines are easily observed on the video recordings and can be seen to be in constant motion, sweeping over the whole surface and moving radially inwards just below the liquid surface, from the peripheral wall through to the centre where they disappear. From features in the lines, their inward radial velocities are in the range 1–10 mm/s.

Combining these Findings with the intermittent cold and hot temperature spikes discovered with the fixed micro-thermometers, it is concluded that the intermittent convection is co-operative in nature. Indeed, the Schlieren pictures show patterns of convection lines which are similar to those observed during extensive studies of the evaporation of water by Rayleigh-Bénard convection [14, 15].

In one of these studies, for example, a shallow bath of water was heated uniformly from below and evolved into a regular array of hexagonal convection or "Bénard" cells (see Fig. 2.8). In each Bénard cell, hot fluid rises in the centre and cold fluid falls

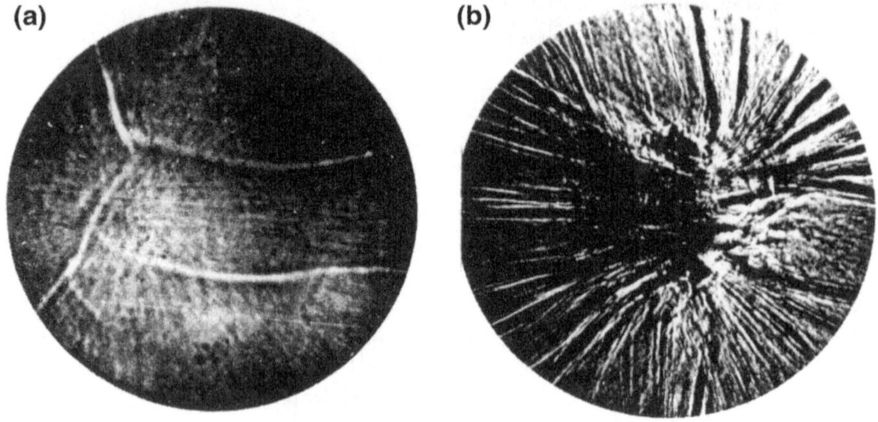


**Fig. 2.7** Schlieren pictures of convection lines in surface of LIN pool, at **a** low evaporation rate, **b** high evaporation rate

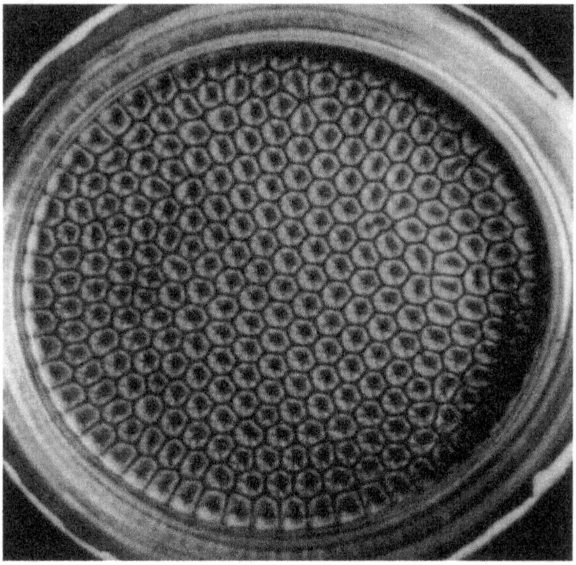

**Fig. 2.8** Laboratory simulation of Rayleigh-Bénard convection in water. Fluid ascends in centre of each cell, and descends along its edge. From Koschmieder and Pallas [17]

along its boundary. The flow within the cell is believed to depend, in fact, on the temperature variation of the surface tension in the free surface. The surface tension decreases with increasing temperature; if an element of the surface is locally hotter than the rest, the liquid is drawn away from the hot spot by the action of surface tension, cools by evaporation and then sinks around the edge of the cell. At the same time, hot liquid is drawn up to the surface through the centre of the cell.

This hexagonal array can also be seen in the following experiment in your kitchen.

Take a flat bottom aluminium saucepan and add 2 cm of water. Bring to the boil and then reduce the heating to a minimum. A stationary hexagonal pattern of bubbles will be seen, each bubble being the centre of a Bénard-type convection cell, despite the uniform heat flux from below, and the extremely small differential surface tension forces in the free surface.

In conventional Bénard convection, the liquid surface is stationary in the laboratory frame of reference, and hence the cells are also stationary. In contrast, the surface of a cryogenic liquid, to a depth well below the surface sub-layer, has a velocity directed radially inwards so that the Bénard cells are drawn out in the moving surface layer into wedge shapes with their boundaries moving radially inward as so-called "streamers".

Each spike in temperature in the vertical temperature profile (of Fig. 2.4) is then due to the passage, past the fixed micro-thermometer, of the edge of a Bénard cell (for a cold spike) and the centre of a cell (for a hot spike).

Another important observation from the video Schlieren recordings is that the number of convection lines, and hence the number of cells enclosed by the convection lines, increases with evaporation rate. Indeed, there appears to be a linear variation between evaporation rate and total length of convection lines or streamers in the surface. The evaporative mass flux per meter length of convection line was never specifically measured, but from the Schlieren video pictures it is estimated to be of the order of 7 g/ms for LIN over a range of surface mass fluxes from 1.0 to 20 $g/m^2$ s.

If one could devise an experiment to scan the surface and measure the local instantaneous evaporation mass flux, one should see a correlation between the passage of a streamer (the edge of a Bénard cell) and a dip in local evaporation mass flux, and surface temperature (at the top of region 2).

Since the streamers are transient in their existence, and variable in number and length with time, it follows that the observed evaporation rate is variable with time, when averaged over integration times of the order of a second. This is particularly so at normal (low) storage evaporation rates where the Schlieren pictures show that only a small length of streamer per square meter of surface is apparently needed to drive the evaporation.

### *2.4.5 The Delicate Evaporation Impedances of the Surface Sub-layer*

Turning Eqs. (2.11) and (2.12) around, we can see that there are three thermal mechanisms which contribute sequentially, by intermittent convection, mixed conduction/convection, and molecular evaporation respectively, to the equilibrium evaporation within a surface sub-layer of about 5 mm thickness.

Firstly, a series of convective Bénard-type cells, with a vertical dimension of about 5 mm, carry superheated liquid up to within about 0.4 mm of the surface through a temperature gradient of the order of 50 K/m. i.e. the convective heat flow is transferred by a strong thermal process with a small impedance.

Secondly, heat is transferred towards the surface, within the 0.4 mm mixed conduction/convection layer, via a very large temperature gradient of the order of 5000–10,000 K/m, by a relatively weak thermal process. With a high thermal impedance, the process consists of a static thermal conductance enhanced about 1.5–2.5 times by penetration of some of the intermittent convection from the Rayleigh/Bénard convection below.

Thirdly, this conducted heat is absorbed by evaporating molecules of the liquid escaping from a thin molecular evaporation region or layer, probably thinner than 1 $\mu$, with a relatively low thermal impedance for a clean surface.

For example, for LIN, and $LCH_4$ respectively, the relative values of the thermal impedance associated with each mechanism are:

- 0.02 and 0.1 for the convection term,
- 250 and 545 for the mixed conduction/convection term and
- 11 and 28 for the molecular evaporation term with $\alpha$ equal to $10^{-3}$.

## 2.5 Surface Sub-layer Agitation and Unstable Evaporation Phenomena

### *2.5.1 Agitation of the Surface Sub-layer*

It can be seen that the large impedance offered by the mixed conduction/convection of the heat flow across region 2, only 0.4 mm thick, is the dominant term.

It is this thin high-impedance region which is separating the bulk superheated liquid from the surface molecular evaporation region, and is preventing a much larger evaporation mass flux from taking place.

Any disturbance or agitation of this thin conduction/convection region, whereby motion of bulk superheated liquid penetrates through it, or replaces it, will result in an immediate, rapid and large, increase in evaporation rate. The surface impedance drops to that of the molecular evaporation impedance and assuming that the evaporation coefficient remains at $10^{-3}$, the evaporation rate rises 23 fold for LIN, and some 19 fold for $LCH_4$, as indicated in Fig. 2.9 [16].

When the disturbance ceases, the surface sub-layer regions re-establish themselves, the high thermal impedance reappears and the evaporation rate falls. Since region 2, the conduction/convection layer, is so thin, the self-repair takes place in a few seconds or minutes and the evaporation rate consequently recovers its previous "normal" value in the same time.

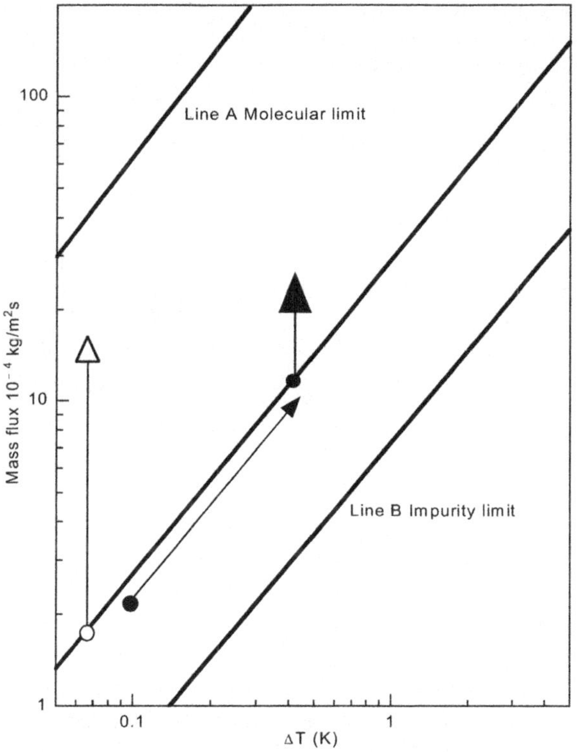


**Fig. 2.9** Variation of evaporative mass flux from LIN with bulk superheat ΔT Line A is upper limit set by molecular evaporation with $\alpha = 10^{-3}$. Line B is lower limit set by surface impurity with $\alpha^{-5}$. ●–●–▲ depicts equilibrium–equilibrium–Mode 1 rollover mass fluxes in Fig. 5.5. ○–Δ depicts equilibrium–Mode 2 rollover mass fluxes in Fig. 5.6

## 2.5.2 *Continuous Irregular and Intermittent Boil-Off*

A number of phenomena have been observed in which agitation modifies the structure of the surface sub-layer and thereby changes the evaporation rate.

On a short time scale, surface evaporation is observed to vary continuously with time, with peaks and troughs in the evaporation mass flux, and hence the surface evaporation impedance, up to 10% of the mean figure. This was the first discovery made on commencing the monitoring of boil-off rates using instantaneously recording vapour flow meters, with an integration time of less than one second.

This observation was the first indication that evaporation is a dynamic process associated with intermittent convection of heat and mass through the surface.

## 2.5.3 *Vapour Explosions*

When the liquid is disturbed by, for example, the transmission of building vibration, or by accidentally knocking the liquid container, the evaporation rate rises rapidly to a high value, and then drops back again just as quickly. The surface evaporation

**Fig. 2.10** Typical time variation of evaporation rate during a vapour explosion

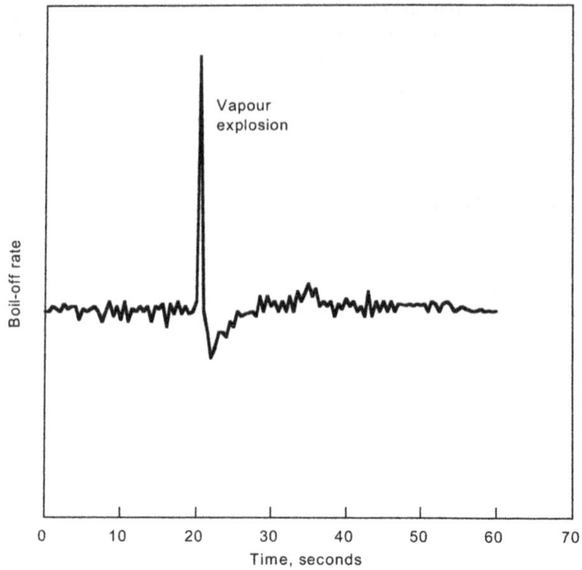


impedance mechanism has broken down quite suddenly and then rapidly repairs itself (see Fig. 2.10).

This evaporation spike could also be reproduced repeatedly by tapping the liquid container at fairly lengthy intervals. Frequently, the flow meter went off scale during the spike, but rapidly came back on scale towards the previous reading. Rocking the container so as to cause the liquid to slosh, would also create a boil-off spike.

These boil-off events are called "Vapour explosions" and demonstrate how sensitive the evaporation impedance of the surface is to disturbance.

We have found that a demonstration Dewar, with a narrow vapour exit into the atmosphere and containing some 5 l of LIN, exhibits "puffs" of vapour, equivalent to mini-vapour explosions, all the time. We have frequently demonstrated this phenomenon in lectures, and it works most obligingly to confirm the irregular surface evaporation of cryogenic liquids.

Vapour explosions are, of course, alarming, but they are surface evaporation phenomena and are generally not dangerous when they arise from transient disturbances of the surface layers.

### *2.5.4 Rollover and Nucleate Boiling Hot Spots*

It is possible for agitation of the surface sub-layer to be maintained continuously over a longer period of time, several seconds, or minutes, or hours in length, so as to prevent the self-repairing mechanism re-establishing the equilibrium sub-layer structure.

The high evaporation rate will then be maintained until the thermal overfill energy, or superheat, of the whole volume of bulk liquid is dissipated by the latent heat of the evaporated mass flow.

Large diameter vents are needed to remove the large volumes of vapour produced and to avoid damage arising from the high pressures that would otherwise be generated.

Examples of continuous agitation of the surface sub-layer include:

(i) the spontaneous mixing between stratified layers of different composition and temperature, called rollover, when the characteristic penetrative convection loops reach the surface and
(j) nucleate boiling from local hot spots whereby a stream of bubbles rise to the surface to disturb and break up the surface sub-layer.

Rollover, as the inevitable consequence of stratification, is more fully discussed in Chap. 5 on multi-component liquids.

### 2.5.5 *QHN Boiling and Geysering*

Vapour explosions should not be confused with QHN boiling, or geysering. In brief, vapour explosions arise from agitation of the surface sub-layer, temporarily reducing the impedance to the evaporative mass flow. On the other hand, QHN boiling arises when the whole of the unstable superheated liquid pool or bath becomes full of boiling bubbles, as described above in Sect. 2.2.3, resulting in unpleasant and possibly dangerous venting of cryogenic liquid and vapour mixed together.

Geysering as defined [17] usually occurs in a vertical tube containing saturated liquid which is subjected to a low heat flux. Periodically, a sudden generation of vapour occurs and liquid is expelled from the top of the tube. Vapour bubbles, formed by heterogeneous nucleate boiling at the walls of the tube, coalesce into large Taylor bubbles filling the cross-section of the tube. These Taylor bubbles then carry liquid up the tube, causing it to superheat and then boil by a process similar to QHN boiling, thereby creating a high velocity jet of liquid and vapour mixture.

## 2.6 Summary of Evaporation Processes

1. In a reasonably well-insulated cryogenic liquid storage vessel, all the heat inflow through the insulation to the roof, walls and base of the vessel is carried to the free liquid surface by convection processes.
2. The evaporation mass flux, in terms of mass evaporated per unit area of liquid free surface, varies with bulk liquid superheat $\Delta T$ as $\Delta T^{1.33}$ for LNG and as $\Delta T^{1.8}$ for LIN and LOX.

3. Conversely, the liquid must be superheated for surface evaporation to take place.
4. Surface evaporation is the only liquid/vapour phase transition taking place, whereby all the heat inflow is absorbed by the latent heat of vaporisation. The enthalpy of the vapour carried out of the vessel balances this heat inflow under equilibrium conditions of storage.
5. Evaporation is controlled by molecular evaporation, mixed thermal conduction/ convection, and Rayleigh-Bénard type cellular convection in successively deeper, identifiable, horizontal thin liquid layers, all within a surface sub-layer of about 5 mm below the vapour/liquid interface. The cellular convection is probably driven by surface tension variations in the free surface of the liquid arising from temperature variations created by local differential evaporative mass flows.
6. During normal equilibrium evaporation, these three mechanisms are in a state of delicate balance, and the evaporation mass flux is determined by the overall temperature difference across these layers, which is the same as the liquid superheat.
7. The three mechanisms act together as a large, but delicate, impedance between the thermodynamic states of the bulk superheated liquid and the saturated vapour above the surface of the liquid.
8. Any disturbance or agitation of the delicate balance between the three mechanisms will generally lead to a 23 fold increase in evaporation rate for LIN and a 19 fold increase for $LCH_4$, as a vapour explosion over a short time or as a rollover event sustained over a longer period. These increases assume that the molecular evaporation coefficient remains at $10^{-3}$ for both LIN and LNG. If $\alpha$ is 10 times larger, the increase in evaporation rates during a vapour explosion, or rollover, will be 10 times larger, i.e. 230 and 190 fold respectively.
9. The three mechanisms appear to be self-repairing after the disturbance, or agitation of the surface sub-layer, ceases.

## References


1. Collier, J.G.: Convective Boiling and Condensation. Oxford University Press, UK (1972)
2. Ashworth, S.P., Beduz, C., Harrison, K., Lavin, T., Pasek, A.D., Scurlock, R.G.: The effect of coating thickness and material on a porous enhanced boiling surface in cryogenic liquids. In: Proceedings of LTEC, Southampton 90, 11.2 (1990)
3. Van Sciver, S.W.: Helium Cryogenics. Plenum Press, New York (1986)
4. Beduz, C., Scurlock, R.G.: Improvements in boiling heat transfer in cryogenic plant: model of co-operation between industry and university. In: Proceedings of ICEC12, Southampton, 319 (1988)
5. Aitken,W.H., Beduz, C., Scurlock, R.G.: The mismatch between laboratory boiling heat transfer data and industrial requirements. In: Proceedings of ICEC15, Genoa (1994)
6. Beduz, C., Rebiai, R., Scurlock, R.G.: Evaporation instabilities in cryogenic liquids and the solution of water and $CO_2$ in liquid nitrogen. In: Proceedings of ICEC9, Kobe, 802 (1982)
7. Tritton, D.J.: Physical Fluid Dynamics. Van Nostrand Reinhold, New York (1988)

8. McAdams, W.H.: Heat Transmission. McGraw Hill, New york (1954)
9. Atkinson, M.C.M., Beduz, C., Rebiai, R., Scurlock R.G.: Heat and evaporation mass transfer correlation at the liquid/vapour interface of cryogenic liquids. In: Proceedings of ICEC10, Helsinki, 95 (1984)
10. Rebiai R.: Solubility of non-volatile impurities in cryogenic liquids. Ph.D. thesis, Southampton University (1985)
11. Atkinson, M.C.M.: Cryogenic Liquid/Vapour and Liquid/Liquid inter facial mass transfer. Ph.D. thesis, Southampton University (1989)
12. Agbabi, T., Atkinson, M.C.M., Beduz, C., Scurlock, R.G.: Convection processes during heat and mass transfer across liquid/vapour interfaces in cryogenic systems. In: Proceedings of ICEC11, Berlin, 627 (1986)
13. Robinson, I.S.: Satellite Oceanography. Ellis Horwood, UK (1995)
14. Davies, J.T., Rideal, E.K.: Interfacial Phenomena. Academic Press, UK (1966)
15. Beduz, C., Scurlock, R.G.: Evaporation mechanisms and instabilities in cryogenic liquids such as LNG. Adv. Cryog. Eng. **39**, 1013 (1994)
16. Bénard, H.: The cellular whirlpools in a liquid sheet transporting heat by convection in a permanent regime. Ann. Chim. Phys. **23**, 62 (1901)
17. Koschmieder, E., Pallas, S.: Heat transfer through a shallow, horizontal convecting fluid layer. Int. J. Heat and Mass Transfer **17**, 991 (1974)
18. Hands, B.A.: Cryogenic Engineering. Academic Press, UK (1986)

# Chapter 3
# Heat Flows into a Cryogenic Storage System: Overall Picture

## 3.1 No Boiling

For any reasonably well-insulated vessel or tank, the major heat flows entering the liquid are via a mix of radiation, conduction and convection to the wetted area of the inner containing wall and to the surface of the liquid. These heat flows include:

- conduction through the insulation space around the inner containing wall,
- convection and radiation within the insulation space,
- conduction down the unwetted neck or container walls,
- convection in the vapour above the liquid, and
- radiation from warmer parts of the container, the roof, the neck, the unwetted walls and through the pipework.

The sum total of the resultant heat fluxes are typically less than 100 W/m$^2$ for LIN, and 100 mW/m$^2$ for LHe.

Following the discussions in Chap. 2, Sect. 2.3, it needs to be restated that the heat flux levels are far too small for nucleate boiling in typically:

- LIN, where the level of heat flux through the insulation of 100 W/m$^2$ is too small by two orders of magnitude, compared with the minimum required to initiate nucleate boiling, via a local heat flux of 10,000 W/m$^2$, producing a wall-liquid temperature difference of about 1 K.
- LHe, where the level of heat flux through typical insulation of 100 mW/m$^2$ is too small, again by two orders of magnitude, compared with the minimum local heat flux of 10 W/m$^2$, producing a wall-liquid temperature difference of around 500 mK which is needed to initiate nucleate boiling.

The starting point for discussing the various heat fluxes is therefore the safe assumption that there is no boiling inside a cryogenic storage vessel or tank. If you cannot believe this assumption, then take a look inside a normally insulated storage vessel using whatever optics or fibre-optics are available. There will be no rising

 
T. D. Bostock and R. G. Scurlock, *Low-Loss Storage and Handling of Cryogenic Liquids*, International Cryogenics Monograph Series,
https://doi.org/10.1007/978-3-030-10641-6_3

streams of vapour bubbles visible whatsoever; only upward convective motion of suspended solid particles of impurities close to the container wall and via thermals at some distance from the wall.

## 3.2 Overall Convective Circulation in the Liquid

With the total absence of any boiling, all heat entering the liquid, by radiation, conduction or convection, is absorbed by primary convection currents which carry the heated (strictly superheated with respect to $T_0$) liquid to the surface, in an open loop circulation (Fig. 3.1).

At the vertical walls, in vessels with depth/diameter ratios of unity or greater, a relatively high velocity, boundary layer flow develops and carries liquid heated by the wall to the surface. At the floor of the vessel, heated liquid in contact with the floor, which has itself been heated from below through the floor insulation, is swept across to join the vertical boundary flow at the walls. The liquid core does not take part in this primary convection.

At the surface, the superheated wall boundary layer flow turns through 90° and moves radially inwards. During this inward radial flow, evaporation takes place as described in Chap. 2.

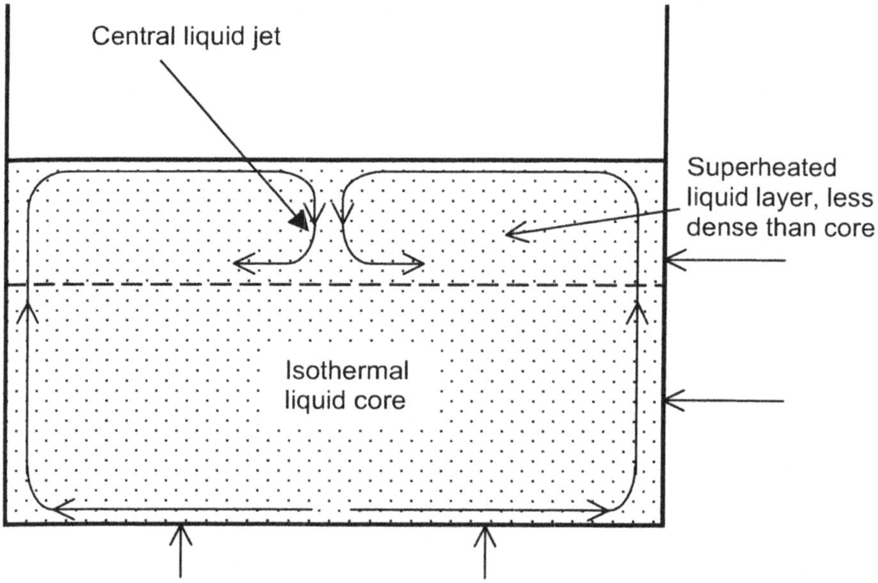


**Fig. 3.1** Overall open-loop convection circulation in liquid producing superheated layer above isothermal core

At the centre, the liquid motion becomes focussed into a strong downward jet. This central jet carries excess superheat, which has not been released by surface evaporation, into the liquid core. Secondary convective processes now produce mixing and distribution of the excess superheat either throughout the core, or alternatively in a stratified layer just below the surface. A large depth/diameter ratio will tend to encourage this type of stratification.

For smaller depth/diameter ratios of the order of 0.5 or less (for example, in large diameter LNG and LPG tanks) additional convection by a number of upward thermal plumes in the liquid will carry heat from the floor to the surface. Each thermal is surrounded by a shell of sinking, colder liquid, the two flows constituting a convective cell, rather like a large Rayleigh-Bénard cell, with a diameter approximating to the liquid depth. Similar, much larger, convective cells occur widely in the atmosphere, and the central rising thermals are used by glider pilots to gain height as standard practice.

From measurements of the local heat transfer from heated vertical surfaces to cryogens in the natural convection regime, it appears that the laminar boundary layer flow undergoes a transition to turbulence when the modified Grashof number ($Gr^*$) is of the order of $10^{13}$. This agrees with experiments performed with water. For a heat flux of 100 W/m$^2$, the wall boundary layer can be expected to be turbulent above a liquid height of 0.3 m in liquid nitrogen, with an increase in heat transfer coefficient for wall/liquid heat transfer [1].

The highest liquid temperature, or maximum liquid superheat, is undoubtedly in the boundary flow at the container wall/liquid-vapour interface. Surface instabilities are likely to be induced near this region where the evaporation mass flux is significantly larger than at the centre of the liquid pool.

Again, it should be mentioned that, since there is no nucleate boiling, the commonly used term "boil-off" is a short and incorrect description of the liquid evaporation. However, it is in common use so we shall use it from time to time in this book.

The "correct" definition of "boil-off" is total evaporative mass rate, in units of kg/s, kg/h or kg/day, or total evaporative liquid volume rate, in units of m$^3$/s, m$^3$/h or m$^3$/day, under specified (assumed) conditions, or the total integrated evaporative surface mass flux, in units of kg/m$^2$ s.

An alternative term is the "percentage boil-off/day", which may be the percentage ratio of evaporative mass rate to storage mass, or of evaporative liquid volume rate to liquid storage volume of the full container.

## 3.3 Thermal Overfill: General Concept

Before discussing the details of heat flows into a cryogenic storage system and the methods for reducing them to acceptable low levels, this chapter starts with building a picture of the overall thermal concept, the absorption of heat flows and their subsequent release by surface evaporation.

In Chap. 2, we showed that the evaporation is driven by part or all of the stored liquid becoming superheated; the greater the surface liquid superheat, the higher the evaporation rate.

The overall amount of superheat energy, or "thermal overfill (TO)", is therefore an important parameter for describing the overall thermodynamic state of the stored liquid. The manner in which TO varies with time is a measure of the stability of the storage system [2].

In formal terms, thermal overfill is the sum of the excess enthalpy $(H - H_0)$ of the stored liquid in relation to the value of $H_0$ defined for the surface of a homogeneous liquid in thermodynamic equilibrium at $T_0$, with its saturated vapour at a prescribed pressure $P_0$. For normal isobaric storage under atmospheric pressure at sea level, the prescribed reference pressure will be close to, but not exactly equal to, 1 bar, and will depend on operational and environmental conditions.

When $(H - H_0)$ is positive for a liquid element, the liquid is described as being "superheated"; when $(H - H_0)$ is negative, the term "sub-cooled" is applicable.

For a large tank, the thermal overfill is defined by:

$$(TO)_{av} = \Sigma(H - H_0) \tag{3.1}$$

where the summation is over all elements of the stored liquid.

Since $(H - Ho)$ is a function of temperature, density, hydrostatic pressure, composition and thermal history, it can be positive or negative for different elements in the same vessel or tank, in which case,

$$(TO)_{av} = (TO)_{+} + (TO)_{-} \tag{3.2}$$

where

$(TO)_+$ is the sum of the positive contributions, and
$(TO)_-$ is the sum of the negative contributions, with a negative sign.

There is frequently poor mixing between different elements of liquid and it is unrealistic to assume that the positive and negative components of $\sum(H - H_0)$ cancel each other out.

The important part, $(TO)_+$, is the positive excess energy in the vessel which, if released by the uncontrolled vaporisation of liquid, could lead to an overpressure in the vessel. It should therefore be borne in mind that, when the term thermal overfill is used, the relevant quantity is usually $(TO)_-$ because of the poor mixing.

Before discussing the concept of thermal overfill, let us consider the basic energy equation relating the heat absorbed by evaporation, $m_{(ev)}\ \lambda$, (where $m_{(ev)}$ is the evaporation mass flow and $\lambda$ is the latent heat of vaporisation) to the total heat inflow Q through the insulation into the liquid.

Then, under equilibrium conditions, the following energy equation applies:

$$Q - m_{(ev)}\lambda = 0 \tag{3.3}$$

Thermal overfill is largely a characteristic describing the liquid "core" which does not take part in the primary convection flow, the latter being driven by heat absorption at the wall and floor, and evaporation at the surface.

Thermal overfill is also time dependent. If the total heat flow into the stored liquid exceeds the heat absorbed by the "boil-off" vapour mass flow rate, $m_{(ev)}$ via the latent heat of vaporisation $\lambda$, then the rate of change of thermal overfill with time is:

$$d(TO)_{av}/dt = Q - m_{(ev)}\lambda \tag{3.4}$$

For safe storage, the left-hand side of all three Eqs. (3.1), (3.3), and (3.4) should be zero.

If the left-hand side of Eq. (3.4) only is zero, then a meta-stable state with constant $(TO)_{av}$, or constant superheat, exists. This is the normal storage situation because, as discussed in Chap. 2, a finite, constant superheat or thermal overfill is a necessary condition accompanying the equilibrium evaporation of a cryogenic liquid.

However, when $d(TO)_{av}/dt$ is positive over a period of time, then a hazardous storage situation is building up; part of the heat inflow is being stored in the liquid and is not being absorbed by evaporation and removed from the liquid. It should be noted that $d(TO)_{av}/dt$ is positive when the atmospheric pressure (and the reference pressure $P_0$) is falling.

The magnitude of the thermal overfill can be very considerable. For example, consider LNG in a large storage tank of 100,000 $m^3$ liquid capacity with bulk superheats of 0.1, 0.2 and 0.4 K. Then the associated thermal overfills are 14,700, 29,400 and 58,800 MJ respectively; large quantities of energy to dissipate.

Compared with the chemical energy stored in the liquid as heat of combustion, these thermal overfill energies are, however, relatively small; only a few per cent. On the other hand, they are physical energies which may be more easily released than chemical energy via triggering mechanisms. These triggers may not be so easily identified.

With a design boil-off rate of 0.03%/day of LNG, i.e. 30 $m^3$ of liquid per day, the equilibrium heat flow into the tank is 75 kW, or 6500 MJ/day.

Using Eq. (2.4), the required superheat is only 0.063 K to generate the necessary surface mass flux of 0.074 $g/m^2$ s to absorb 6500 MJ/day at an equilibrium thermal overfill of 8800 MJ.

In the example, the LNG had bulk liquid superheats of 0.1, 0.2 and 0.4 K, with corresponding evaporative mass fluxes of 0.12, 0.26 and 0.57 $g/m^2$ s (from Eq. 2.4) which are equivalent to boil-off figures of 0.05, 0.1 and 0.23%/day. These boil-off rates will dissipate thermal overfill at the rates of about 10,500, 22,750 and 49,800 MJ/day, daily figures which are similar in magnitude to the thermal overfills of the bulk superheated liquid. It therefore follows that the excess thermal overfill from bulk superheats of 0.1, 0.2 and 0.4 K, will be dissipated by excess boil-off, falling from peaks of about 1.6, 3.5 and 7.6 times the normal rate respectively, back to the normal rate over a period of about one or two days.

## 3.4 Distinction Between 'A' and 'B' Heat In-Flows

It is important to distinguish between heat flows through the liquid surface and the wetted walls and floor, the 'A' flows; and those through the unwetted walls, the 'B' flows, of the storage vessel (see Fig. 3.2).

The 'A' heat flows are absorbed by the latent heat of vaporised liquid together with superheating of the bulk liquid as thermal overfill. On the other hand, most or all of the 'B' heat flows can generally be absorbed by heating the vapour only, i.e. by increasing the sensible heat or enthalpy of the cold vapour between boiling point and ambient temperature.

With good design, the 'B' heat flows may not contribute to the evaporation at all, although they may rely on a minimum vapour mass flow generated by 'A' heat flows.

Distinction between the absorption of 'A' and 'B' heat in-flows for different cryogenic liquids can be made by reference to Table 3.1.

The ratio of available sensible heat/latent heat varies from 75.5 for helium, down to 0.25 for propane. In general, if the ratio is less than about 0.5, the vapour cooling effect becomes insignificant, and this is so for krypton, ethane and the higher hydrocarbons.

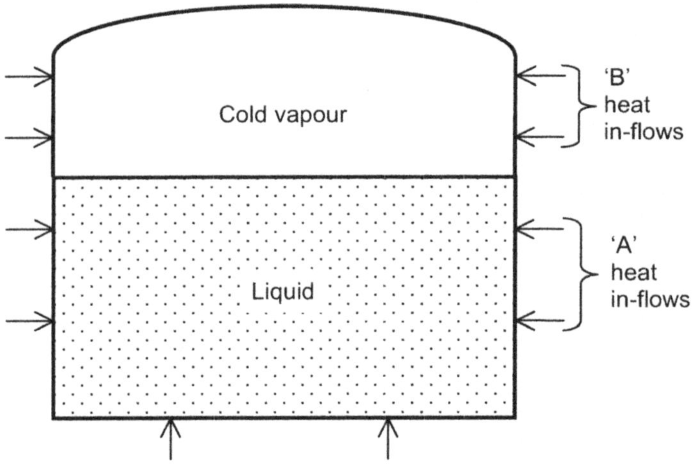


**Fig. 3.2** Distinction between 'A' heat in-flows absorbed by liquid evaporation and 'B' heat in-flows absorbed by cold vapour

**Table 3.1** Comparison of latent heat and sensible heat of cryogens

| | $T_s$ (K) | $\rho$ (kg/m$^3$) | $\lambda$ (kJ/kg) | $\Delta H_s$ (kJ/kg) | $\Delta H_s/\lambda$ |
|---|---|---|---|---|---|
| Helium 4 | 4.22 | 125 | 20.72 | $1564^{4\text{–}300K}$ ($422^{4\text{–}80K}$) | 75.5 (20.37) |
| n-Hydrogen | 20.28 | 70.8 | 445.45 | $3511^{20\text{–}300K}$ ($209^{20\text{–}80K}$) | 7.88 (2.04) |
| Neon | 27.09 | 1205.2 | 85.71 | 282.9 | 3.30 |
| Nitrogen | 77.31 | 806.79 | 198.86 | 234.5 | 1.18 |
| Air | 78.9 | 875 | 202.6 | 222.3 | 1.10 |
| Argon | 87.28 | 1394 | 161.34 | 112.8 | 0.70 |
| Oxygen | 90.19 | 1141.2 | 212.46 | 193.1 | 0.91 |
| Methane | 111.69 | 422.4 | 510.33 | 404.0 | 0.79 |
| Krypton | 119.77 | 2413.9 | 107.93 | 45.7 | 0.42 |
| Ethylene | 169.41 | 568.0 | 482.58 | 182.0 | 0.38 |
| Ethane | 184.55 | 544.1 | 488.49 | 162.0 | 0.33 |
| Carbon dioxide | 194.7** | 1560 | 563 | 82.0 | 0.15 |
| Propane | 231.1 | 581 | 426 | 107.0 | 0.25 |
| n-Butane | 272.6 | 601 | 386 | 40.5 | 0.10 |

**Sublimation temperature

Thermally efficient design of a storage vessel consists, therefore, of reducing the 'A' heat in-flows to a minimum, while ensuring that the 'B' heat in-flows are absorbed by the cold vapour and do not enter the liquid. Furthermore, any conversion of 'A' heat in-flows into 'B' heat in-flows will be most beneficial in further reducing the liquid evaporation rate, a trick which can be achieved in several ways described below.

Let us now outline the various heat flows by radiation, conduction and convection respectively, distinguishing between the 'A' heat in-flows and the 'B' heat in-flows. The next chapter (Chap. 4), will go on to describe, in detail, the techniques available to us for controlling and reducing the heat flows.

## 3.5 Radiative Heat In-Flows

Cryogenic liquids and vapours are generally transparent to infra-red radiation with wavelengths greater than 1 μm, i.e. to nearly all black-body radiation from 300 K and colder sources. In contrast, all plastics, glasses and oxidised surfaces are totally opaque to infra-red radiation at these wavelengths, with reflectivities close to zero (and emissivities close to unity). On the other hand, metals may have very high reflectivities, and emissivities as low as 0.01 (for Au and Ag), ~0.02 (for Al) and ~0.05 to 0.10 for polished steels.

These low emissivity values for metals can be used advantageously to reduce radiation fluxes as will be described later, but they may rise rapidly and approach unity for oxidised or rough metal surfaces.

Ambient temperature 300 K radiation incident on the liquid surface, an 'A' heat in-flow, will be transmitted through the liquid to, and absorbed at, the inner walls and base of the enclosing vessel. This is perhaps surprising, but with cryogenic liquids having total transparency to infra-red radiation around the 300 K peak, they do not absorb any energy from the radiation.

'B' in-flow radiation incident on the unwetted walls will be totally absorbed if the walls are rough and have a high emissivity of unity. If the walls are smooth and have a low emissivity, then the incident radiation will be reflected down into the liquid as an 'A' heat in-flow.

Radiation in pipework may be a problem since it can be funnelled via reflections at grazing incidence into the liquid as 'A' heat in-flow, with little absorption on the way.

## 3.6 Conductive Heat In-Flows

The main 'A' heat in-flow of conducted heat is through the insulation in the space surrounding the inner liquid vessel. Other sources include conduction down the neck wall, through the mechanical supports of the inner vessel, and through the pipework. Conduction through the vapour in contact with the liquid is generally small, but should not be forgotten, particularly when the liquid has a large surface area, and when the vapours have relatively high thermal conductivities, like helium and hydrogen.

In very large tanks, insulation below the floor of the tank has to be load-bearing across the whole floor area, and techniques to prevent freezing of the underlying ground have to be used to avoid the serious problem of frost-heave. As a result, the 'A' heat in-flow through the floor insulation tends to be the dominant conduction term.

'B' heat in-flows include conduction down the neck or tank wall from the top, and through the insulation to the unwetted walls; also conduction along pipework entering the liquid from the top plate or roof.

## 3.7 Convective Heat In-Flows

Density stratification in the vapour space over the liquid, would be expected to act so as to prevent convection. In practice, however, there is considerable convective motion of the vapour (see Fig. 3.3).

Heating of the vapour by the wall generates a strong natural convective boundary layer flow up the wall. At the same time, there is a reverse flow in the centre of the vapour column which carries heat as an 'A' in-flow to the surface of the liquid.

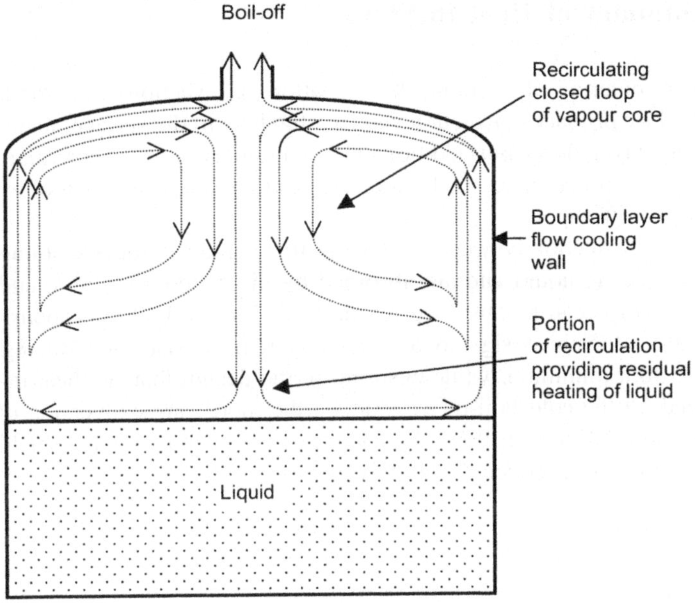


**Fig. 3.3** Convective circulation of vapour with recirculating closed loop of vapour core, a portion of which provides residual heating of the liquid

The wall boundary layer flow in turn ensures good cooling of the walls by the cold vapour. This vapour cooling effect is a bonus since it can be used to absorb, as 'B' heat in-flows, part or all of the conducted heat flows down the walls which would otherwise enter the liquid as 'A' heat in-flows.

The heat transfer in the steep vertical temperature gradient in the vapour space is also greatly enhanced, and this aids the efficiency of the vapour cooling effect. The purpose of most insulation material with a cellular or fibrous structure is to restrict gas convection within it. When voids and large cells occur, then strong convective circulation takes place, with large heat flows across the cells.

## 3.8 Other Sources of Heat Flow into the Liquid

The most important additional source of heat flow is that arising from thermoacoustic oscillations in the vapour columns above the liquid which can lead to very large 'A' heat flows into the liquid.

At liquid helium temperatures, and increasingly so at decreasing temperatures below 1 K and into the milli-Kelvin region, other heat in-flows become important. These include mechanical vibrations, eddy current heating and the Joule heating of measuring instruments.

## 3.9 Summary of Heat In-Flows

1. All heat entering the liquid, by radiation, convection and conduction, is absorbed by primary convection carrying the heat to the surface, mainly by wall boundary layer flows but additionally by thermals in low depth/diameter ratio tanks, where it may be carried out of the liquid by the latent heat of evaporation of the boil-off vapour.
2. There is a distinction between 'A' heat in-flows into a cryogenic storage system, which enter the liquid and are absorbed by the evaporation loss, and 'B' heat in-flows which can be absorbed by the "cold" in the boil-off vapour.
3. Thermally efficient design of a storage system consists of reducing 'A' heat in-flows to a minimum, while ensuring in the design that 'B' heat in-flows are absorbed by the cold boil-off vapour. Furthermore, any conversion of 'A' heat in-flows into 'B' heat in-flows, by the use of, say, vapour cooled shields, will be most beneficial in reducing liquid evaporation.

## References

1. Beduz, C., Rebiai, R., Scurlock, R.G.: Thermal overfill and the surface evaporation of cryogenic liquids under storage. Adv. Cryog. Eng. **29**, 795 (1983)
2. Beduz, C., Rebiai, R., Scurlock, R.G.: Evaporation instabilities in cryogenic liquids and the solubilities of water and $CO_2$ in LIN. In: Proceedings of ICEC9 (Kobe), 802 (1982)

# Chapter 4
# Insulation: The Reduction of 'A' and 'B' Heat In-Flows

The distributed heat flows, through the unwetted walls into the vapour, are part of the B heat flows. They cause a convective vertical boundary layer flow up the walls which can absorb the conducted heat down the walls via a distributed heat transfer. The net conducted heat flow into the liquid can be reduced to zero with the right design.

The second part of the B flow is the ambient temperature infra-red radiation down the neck or from the container roof. This can be reduced or eliminated, with one or more vapour cooled horizontal shields in the cold vapour. Again this heat flow is reduced via a distributed heat transfer, this time by conduction across to the vapour boundary layer.

The A flows into the liquid through the wetted wall and floor of the storage container or tank cause a convective vertical boundary layer flow at the walls. This boundary layer flow rises to the liquid surface where the complex, but delicate, evaporation process takes place across the whole surface area.

## 4.1 Reduction and Control of Heat In-Flows

To minimise heat inflow, it is important to identify all heat sources and then to reduce all their heat inflows to the same order of magnitude, or to zero if practicable. It is also important to recognise that variation with scale is different for the various heat sources and therefore the effort and cost of reducing the heat in-flows will vary with scale.

The first indication of poor insulation is the presence of snow or ice on the outside of a cryogenic system. This should not be tolerated.

The presence of snow on a vapour vent line is also an indication of a poor design which is not making adequate use of the cold vapour in absorbing heat flows.

A number of techniques are available for converting 'A' heat in-flows into 'B' in-flows, and these will be outlined below. The most important include:



T. D. Bostock and R. G. Scurlock, *Low-Loss Storage and Handling of Cryogenic Liquids*, International Cryogenics Monograph Series,
https://doi.org/10.1007/978-3-030-10641-6_4

- Vapour cooled radiation baffles.
- Vapour cooling of the unwetted neck wall to reduce neck conduction.
- Vapour cooled multi-shields to reduce conduction through the insulation space.

## 4.2 Radiation

### *4.2.1 Stefan's Law and Low Emissivity Materials*

The major source of infra-red radiation into the liquid is the warm upper parts of the storage vessel, and this heat flow is governed by Stefan's Law, which states that the radiation flux between parallel surfaces at absolute temperatures $T_1$ and $T_2$, with $T_1 > T_2$, is given by:

$$Q/A = \sigma e_1 e_2 \left(T_1^4 - T_2^4\right)/\left(e_1 + (1 - e_1)e_2\right) \quad (4.1)$$

where $e_1$ and $e_2$ are the "total emissivities" of the two surfaces at temperatures $T_1$ and $T_2$ respectively, and Stefan's constant, $\sigma = 5.67 \times 10^{-5}$ kW/m$^2$ K$^4$. For 300 K black body radiation, the radiation peak is at 10 μm wavelength in the infra-red, and Q/A is 460 W/m$^2$ (i.e. for $e_1 = e_2 = 1.0$ and $T_2 \ll T_1$).

Figure 4.1 shows the energy versus wavelength distribution for several cryogenic temperatures. As the temperature falls below 300 K, the total radiant energy (or the area under the energy–wavelength curve) decreases very rapidly, being proportional to the fourth power of the absolute temperature. At the same time, the radiation peak wavelength becomes greater. This peak wavelength, $\lambda_{max}$ is given by Wien's displacement formula:

$$\lambda_{max} T = 2898\ \mu\,°K \quad (4.2)$$

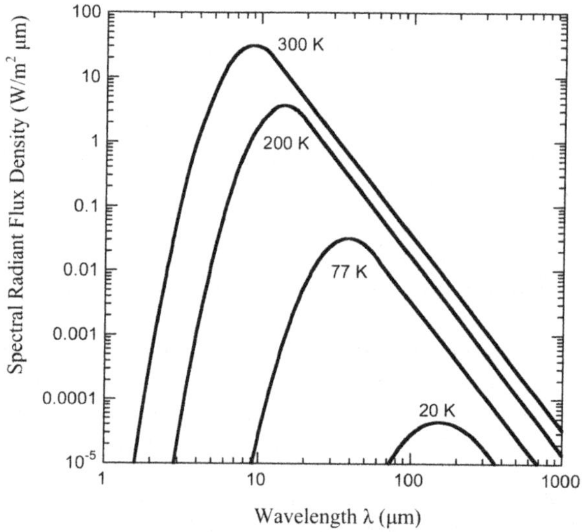


**Fig. 4.1** Blackbody radiation spectra for a number of temperatures

Now, the total emissivity is the integral over all wavelengths of the spectral emissivity weighted according to the black-body energy–wavelength distribution function. For a "grey" body, the total and spectral emissivities are the same. Fortunately, metals approximate to grey bodies reasonably well at the wavelengths that are important for thermal radiators at 300 K or less; that is, for wavelengths greater than a few microns.

This fact enables us to use the expressions "emissivity of the cold surface" and "absorptivity of the cold surface for radiation from the warm surface" as if they were interchangeable. With this in mind, we will continue to use emissivities as though they are independent of both temperature and wavelength.

From the available data on low temperature emissivity materials [1], several generalisations can be made, namely:

- The best reflectors (with the lowest emissivities) are also the best electrical conductors, e.g. silver, e = 0.01, gold, e = 0.01, copper, e = 0.015 and aluminium, e = 0.02 (all reflectors at 77 K and exposed to 300 K black body radiation).
- The emissivity **decreases** significantly with decreasing temperature.
- The low emissivity of good reflectors is **increased** by surface contamination, e.g. grease and oxidation.
- Alloying a good metallic reflector **increases** its emissivity, e.g. aluminium alloys have emissivities of 0.06 and higher.
- The emissivity is **increased** by work-hardening the surface layer, e.g. by mechanical polishing to give a shiny finish.
- Visual appearance is not a reliable guide to reflecting power at infra-red wavelengths of 10–100 μm.

We can now begin to see how the 300 K radiation heat flow can be reduced by a large factor by using low emissivity baffles, cooled by the cold vapour to a temperature of 150 K or less in the neck of the vessel, or by using a suspended deck in the cold vapour above the liquid surface in large tanks.

### 4.2.2 *Vapour-Cooled Radiation Baffles*

As mentioned in Chap. 1, the first systematic studies on the use of vapour-cooled baffles were carried out in the early 1960s at Southampton University, and were reported in 1965 at an IIR conference in Grenoble, France by Lynam, Proctor, and Scurlock [2]. Their paper describes experimental work which showed how ambient temperature radiation, down the neck of wide-necked and short-necked Dewars containing liquid helium, may be almost completely and simply absorbed by positioning a set of two to four vapour-cooled horizontal disc baffles, with a vertical separation of 40–50 mm between each baffle, at the approximate 77 K temperature level in the neck (Fig. 4.2a and b). While gold or silver-plated copper baffles gave the lowest evaporation rate, this rate was only 7.5% lower than with plain copper. Surprisingly, the difference in performance between copper and non-metallic methyl methacrylate (Perspex) baffles was small.

The disc baffles work in the following way. The radiation heat flow down the neck is partially absorbed and partially reflected back up the neck by the baffles, positioned high in the neck (see Fig. 4.3a). The baffles are in turn cooled by increasing the enthalpy of the evaporated helium vapour. For each baffle, the radiative heating is balanced by the vapour cooling. In this way, with a series of baffles, the ambient radiation heat flow is almost completely stopped from entering the liquid and contributing to the liquid evaporation. The radiation, an 'A' heat inflow, has been almost completely converted into a 'B' heat inflow.

We now know that the cooling of the baffles by the rising cold vapour takes place via an efficient heat transfer process, because natural convective heat transfer is greatly enhanced (perhaps as much as 10 fold or more) in the vertical temperature gradients in a Dewar neck.

This simple technique is not confined to the low-loss containment of liquid helium, but is applicable to all cryogenic systems, including small-scale cryocooler refrigerated systems, all storage Dewars, vessels and tanks for cryogenic liquids, and the very large scale storage of LNG, LPG and other hydrocarbon liquids.

In very large LNG tanks, the vapour cooled baffle system is called a suspended deck, which has revolutionised and simplified their design and has, at the same time, significantly reduced construction costs. See Fig. 4.3a and b.

The use of vapour cooled baffles was the subject of several patents back in 1965, but since the technique is so simple and cheap to apply, the general use of baffle systems has spread very quickly into all areas of cryogenics regardless of patents. This general application has therefore led to significant savings in the capital costs of all storage vessels and tanks, and in running costs via reduced liquid boil-off rates and lower refrigeration needs of all cryogenic systems.

There are two simple alternatives to vapour cooled baffles which have been tested, namely plastic foam plugs and floating ball blankets, but both are not so effective.

### *4.2.3 Plastic Foam Plugs*

In the original work on baffles published in 1965 [2], expanded polystyrene or polyurethane foam PUF plugs were demonstrated to be as effective as horizontal baffles and this finding led directly to the widespread use of foam plugs in Dewar necks. However, subsequent work, published in 1969 [3], demonstrated clearly that foam plugs become unreliable and ineffective insulators after continuous exposure to boil-off gas over a few days. See Fig. 4.2b.

In other words, when the low thermal conductivity, high molecular weight, foam gas within the plastic foam is replaced by diffusion by high thermal conductivity, low molecular weight helium or nitrogen gas, the foam plug becomes ineffective,

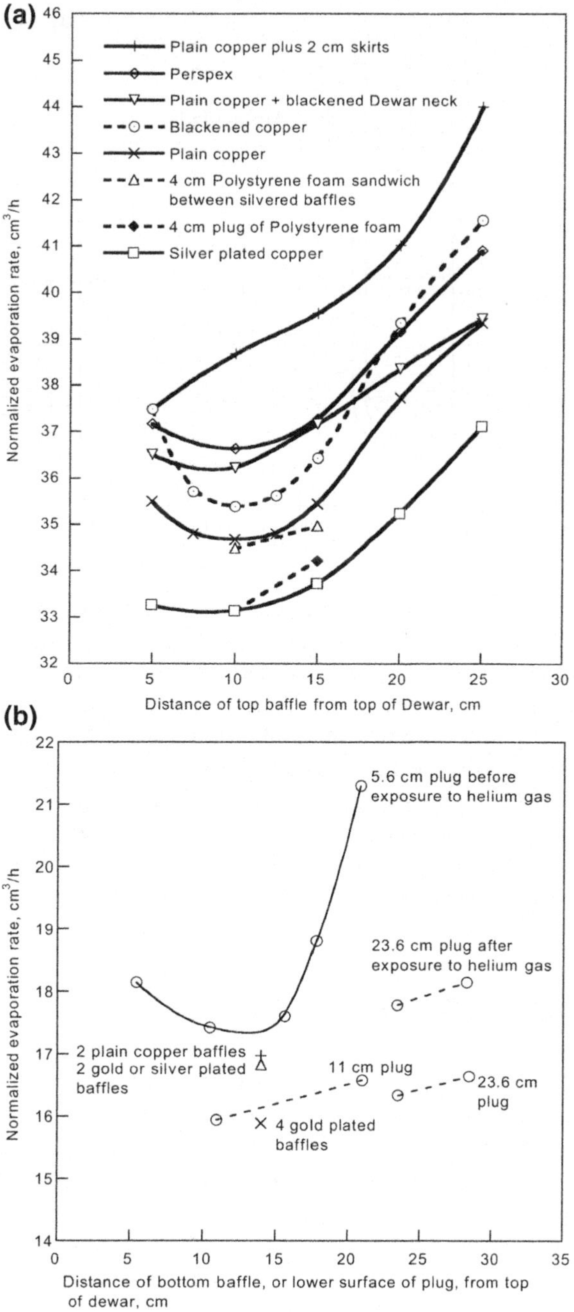


**Fig. 4.2** **a** Variation of LHe evaporation rate with position for various forms of two baffles spaced 4 cm apart. **b** Comparative performance of metal baffles and plugs of polystyrene foam before and after exposure to helium gas

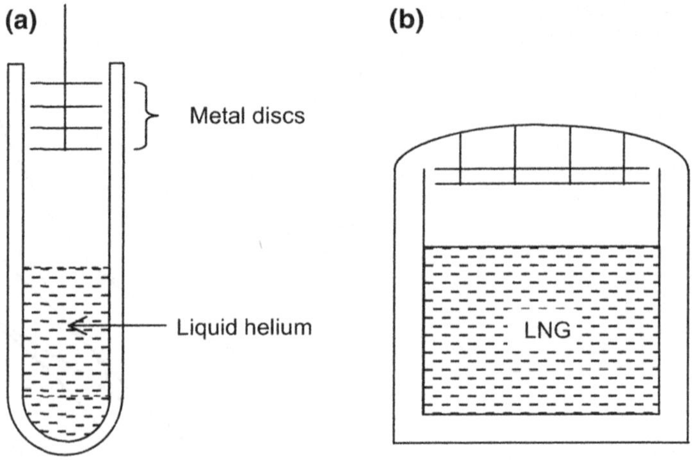


**Fig. 4.3** **a** Vapour-cooled radiation baffles in the neck of a liquid helium Dewar. **b** Vapour-cooled suspended deck in upper section of an LNG storage tank

behaving as a thermal short circuit in the Dewar neck. This thermal short circuiting effect can be easily detected by simply removing the foam plug from the Dewar neck, when the system will have a lower boil-off rate!!

If a set of horizontal metal baffles is, now, placed in the neck, a significantly lower boil-off rate will result, with gold-plated copper yielding the lowest figure. This finding seems to have been forgotten.

The common practice of using PUF plugs in the necks of LIN storage vessels and cryobiological specimen containers may **not** be cost effective. The plugs should be replaced by vertical stacks of 2–4 horizontal metal disc baffles (or even plastic discs) attached to the top cover or lid by low thermal conductivity rods.

### *4.2.4 Floating Ball Blankets*

Hollow plastic balls, 10–20 mm diameter, either plain or aluminised, are commonly used to reduce evaporation of volatile liquids in storage tanks at ambient temperatures. They may also be used to absorb radiation heat flows into cryogenic liquids.

Tests showed that a minimum of 3–5 layers floating on the surface were effective [4]. However, over a long period of time, it is expected that vapour diffusion through the plastic walls will eventually lead to liquid condensing inside the balls. The balls will then sink to the bottom of the tank or, if the layers are stirred up during filling or decanting operations, explode on reaching warmer parts of the tank. These expectations, together with the fact that their plastic composition is brittle at low temperatures, has led to very little application of this simple idea.

Recently there has been renewed development of the idea, applied to reduce sloshing in LNG carriers, and floating LNG (FLNG is the common abbreviation) facilities. Open cell foam cubes, made from melamine resin, encapsulate a buoyant material; these are connected together via a fabric material. This is floated on top of the LNG, and the inertia of the blanket reduces the sloshing of the cargo thereby reducing boil off [5].

## 4.3 Conduction Through the Insulation Space

### *4.3.1 Dewar's Dewar*

Conducted heat, whether by gas conduction, solid state conduction or through solid-solid contacts, is reduced by use of the correct type and thickness of insulating materials in the insulation space surrounding the primary liquid container, vessel or tank.

Sir James Dewar, at the Royal Institution, London, was among the first to realise in 1877 that he needed better insulation to contain the small quantities of cryogenic liquid, he was producing, for a sufficiently long time to make physical measurements on them. He therefore spent his first 15 years at the Royal Institution on developing cryogenic insulations, before he was later able to liquefy hydrogen in 1898.

He concentrated on evacuated double-walled glass vessels into which he placed a variety of powder, paper and solid materials available to him in the 1880s (at that time, perlite and silica aerogel powders, plastics in foam or any other form, and glass fibre did not exist). Most materials did not work for him, including one consisting of three layers of aluminium foil cigarette packaging [6]. Had he doubled the number of layers, he would undoubtedly have discovered the principle of multi-layer insulation (MLI) some 70 years before Peterson's publication in 1951—but he missed that discovery [7].

What he did discover by chance, arose from the mercury-vapour vacuum pump he was using to evacuate his experimental double-walled glass vessels. Some mercury droplets had condensed into the vacuum space, on the inside of the outer glass wall. When the inner vessel was filled with liquid air, a shiny mercury film was progressively deposited on the inner glass container because of the finite vapour pressure of mercury at ambient temperature. The liquid air evaporation rate progressively fell to a lower value than with any of the previous insulations he had tested.

Dewar noted this and, in 1892, cleverly replaced the mercury film with a silver coating to achieve the same low evaporation. At last, after 15 years of systematic study, he had mastered the problem of containing liquid air and later, liquid hydrogen, in an evacuated, double-walled, silvered glass vessel or "Dewar".

In fact, he never patented this discovery, and it was his German glass blower, Herr Müller, who developed the idea after he discovered one night that a "Dewar" vessel was also good for keeping liquid hot (the milk for his baby's feeding bottle during the night!). The Thermos Flasche or Thermos flask was born that night, and very soon went into mass production, in the first place as a German product.

Dewar's Dewar works by cutting out conduction with the use of a vacuum in a low thermal conductivity glass vessel, and reducing radiation by the use of low emissivity reflecting films of silver. It remained the standard laboratory container for cryogenic liquids until the 1960s, and no improvements in evacuated insulations took place before then. The only modification required for liquid use was the substitution of low borosilicate "Monax" glass for the more usual high borosilicate Pyrex glass, owing to the latter's porosity to helium gas at ambient temperature.

### *4.3.2 Gas Purged Insulations at 1 bar*

There is no doubt that vacuum insulation, with the requirement for an additional outer case strong enough to withstand a collapsing pressure differential of 1 bar, is a complicating nuisance which is avoided for economic reasons on large scale industrial applications. It also has to be avoided, because of its weight penalty, for space applications prior to take-off.

Avoiding vacuum insulation would enormously simplify the industrial, and possibly domestic, application of High Temperature Superconductivity in the future. Let us therefore first look at insulations with no vacuum, and then consider briefly whether they might be improved. A wide variety of materials are available today, some natural, but most are synthetic (see Table 4.1).

The basic idea of an insulation at 1 bar pressure is to reduce or eliminate heat transfer by gas convection by the creation of sufficiently small gas cells within a matrix of low thermal conductivity powder, solid fibres or foam walls.

**Table 4.1** Effective thermal conductivities of gas-purged and evacuated insulations between 300 and 77 K, in mW/mK

| | Pressure 1 bar nitrogen | Evacuated to 0.1 Torr |
|---|---|---|
| Expanded perlite | 26–44 | 1.0 |
| Silica aerogel | 19 | 1.6 |
| Fibre glass | 25 | 1.7 |
| Foam glass | 35–52 | – |
| Balsa wood | 49 | – |
| Expanded polystyrene | 24–33 | – |
| Polyurethane foam (PUF) | 25–33 | – |
| Rock wool | 30–43 | – |

At the same time, the insulation is required to be load-bearing, ductile or non-brittle, easy to apply and trouble free in operation, as well as having a thermal contraction to match that of the inner vessel.

To use insulating materials at atmospheric pressure, continuous gas purging, at a small overpressure, is necessary for two reasons:

(1) To avoid the ingress of water vapour condensing and freezing. The effect of liquid water in the insulation above 273 K is serious enough because water has a high k-value, in the region of 550 mW/mK, which is 10–25 times the k-value of the insulation. However, once the water is frozen to ice below 273 K, the insulation efficiency is even more seriously impaired because ice has a much higher thermal conductivity, with a k-value of 2200 mW/mK at 273 K, which is 40–100 times the k-value of the insulation.

Furthermore, if the insulation is open to the atmosphere, the freezing of water vapour and formation of ice is progressive, and water vapour will be sucked into the freezing zone to create ice-bridges, which are literally thermal shunts across the insulation. The use of load bearing, rigid foams, such as PUF and cellular glass, also lightweight perlite concretes, must rigorously include avoiding the ingress of water throughout manufacture, distribution and assembly.

(2) To prevent the ingress of atmospheric air and its partial condensation as oxygen-rich liquid through contact with cold surfaces below 81 K. The build-up of oxygen-rich liquid within the insulation constitutes a safety hazard to be avoided even with non-flammable insulation, besides introducing an additional heat flow via its latent heat of condensation.

The answer to both these problems is to incorporate a water-vapour barrier in the insulation, and use a purge gas at slightly over atmospheric pressure in a gas-tight but not vacuum-tight insulation space. The choice of purge gas is determined by its ready availability and it must obviously not condense at the temperature of the contained liquids.

Dry nitrogen gas purged insulations are commonly used in cold boxes on air separation units, large static storage tanks greater than 3 m diameter, for LOX, LA, and LIN, and in sea-going LNG tankers.

Boil-off gas is commonly used as a purge gas in the insulation of land-based storage tanks for LNG, liquid ethylene and liquid petroleum gases (LPG). In sea-going LPG tankers, propulsion engine exhaust gas may be used as a purge gas, provided the oxygen content remains well below the limiting oxygen index of 11.5 vol.%, i.e. in the region of 2 vol.%.

The outer cladding has to be gas-tight, but since it is under a low pressure differential, it can be built up from thin flat panels to form cold boxes.

In the absence of gas purging in, say, small static systems like refrigeration cabinets, ingress of water is a problem which may be reduced by the use of hydrophobic insulation material, as well as by the use of impermeable vapour barriers as part of the insulation. It should be noted that the relatively low k-value of the foam-gas in foam materials, such as expanded polystyrene or PUF, is soon lost when it is replaced through diffusion by purge gas or atmospheric air with a higher k-value.

The lower limit of heat flux in a gas-purged insulation is determined by the thermal conductivity of the gas within the cells or matrix of the insulating material. For example, the thermal conductivities or k-values of some typical purge gases at 1 bar are given in Table 4.2.

As can be seen by comparing Tables 4.1 and 4.2, the lowest k-values for nitrogen gas purged insulations are in the region of 25 mW/mK, the quoted value for nitrogen gas at 300 K.

It is interesting to consider whether it is possible to obtain lower k-values down to 10 mW/mK at 300 K and how one might achieve this reduction of only 2.5 fold in a purge gas atmosphere at 1 bar. Tests with multi-layers of PUF, or other foam materials, with reflective aluminised Mylar sheets between them, have not seen significantly lower k-values being obtained, so far. However, the idea is very worthwhile pursuing because of the considerable impact it would have, if successful, not only in refrigeration engineering but also in the commercial applications of High Temperature Superconductivity.

For the example of a spherical tank holding 1000 tonnes of LIN (13 m diameter), with 0.2 m thickness of perlite insulation, the heat influx is 16 kW, equivalent to a boil-off of 7 tonnes per day, or 0.7% per day, which is acceptable.

On the other hand, for a spherical tank of 1.3 m diameter holding 1 tonne of LIN, and using the same thickness of perlite, the boil-off will be 7.0% per day, which is not economical; another type of insulation with a lower k-value is needed.

In practice, there are problems with powders, and to some extent with fibre blankets, due to thermal contraction and expansion of the inner vessel and pipework. Over a period of time, the insulations settle or move so as to form voids. The presence of such voids are indicated by the appearance of cold spots or frost patches on the outer containing skin.

**Table 4.2** Thermal conductivities of typical purge gases at different temperatures and 1 bar

| | k (mW/mK) | T (K) |
|---|---|---|
| Nitrogen | 9.6 | 100 |
| | 11.5 | 120 |
| | 25.8 | 300 |
| Helium | 16.9 | 10 |
| | 83.3 | 120 |
| | 156.0 | 300 |
| Methane | 12.8 | 120 |
| | 34.1 | 300 |

### 4.3.3 Evacuated Powder Insulations at 0.1 Torr

By reducing the gas pressure to 0.1 Torr, using mechanical rotary vacuum pumps only, the k-value of powder insulations is reduced by a factor of 10 or more (see Table 4.1). The gas conduction component has been largely removed. However, a penalty has to be paid since the outer casing must now be vacuum tight and strong enough to withstand a collapsing pressure differential of 1 bar. With vacuum insulation, the geometry is limited to cylinders, spheres and combinations for mechanical strength, and the size is limited to about 4 m diameter, above which the weight of the outer vacuum-containing vessel becomes prohibitive.

Thus, for the example of a 1 tonne spherical tank for LIN storage, the boil-off with evacuated perlite insulation will be ~0.7% per day, which is acceptable. Such tanks with evacuated powder insulation are used as customer storage tanks, and on road and rail tankers.

In the latter cases, it is necessary to guard against the settling of powder insulation under mechanical vibration and the creation of voids and cold spots.

### 4.3.4 Multi-layer Reflective Insulations (MLI) at 0.0001 Torr

Evacuated powders do not address the main residual heat flow through this type of insulation, which is infra-red radiation.

Mixing the powder with low emissivity aluminium or copper powder helps to reduce the radiative heat flow by a factor of 4–5, yielding k-values down to 0.4 mW/mK, as was discovered by Kropschot and his colleagues at NBS, Boulder, USA, in the late 1950s [8]. However, it was found that the wide difference in density between powders like expanded perlite and metal particles of aluminium or copper could lead to separation caused by vibration and packing, with the progressive development of thermal shunts within the powder mixture.

The real breakthrough was made by Peterson in his thesis "The Heat-Tight Vessel" in 1951, when he described replacing the powder with a number of thermally isolated reflecting layers of reflective material, or Multi-Layer Insulation (MLI) [7].

However, before discussing MLI it must be born in mind that, below a pressure of about 0.1 Torr, the thermal conductivity of a gas is no longer independent of pressure but decreases linearly with decreasing pressure—the Knudsen flow or free molecular flow region. This means that the pressure operating between the multiple reflectors must be lower than 0.0001 Torr for the gas conduction heat flow to be less than the radiation heat flow through the MLI. If this condition is not met, then gas conduction effects will dominate and the effectiveness of the multi-layers will be lost.

For n thermally floating reflecting surfaces, with emissivity e, the radiation heat transfer between parallel surfaces at temperatures $T_1$ and $T_2$ is given by:

$$Q/A = e\,\sigma\left(T_1^4 - T_2^4\right)/2(n+1)\ \text{kW/m}^2 \tag{4.3}$$

where $\sigma = 5.67 \times 10^{-5}$ kW/m$^2$ K$^4$.

For standard conditions of temperature across the insulation. 300–77, 300–20, 300–4.2 K, an effective mean thermal conductivity $k_{eff}$ can be defined using the standard conductivity equation:

$$Q/A = k_{eff}(T_1 - T_2)/t\ \text{kW/m}^2 \tag{4.4}$$

where t is the thickness of the insulation, not necessarily the width of the insulation space.

Combining Eqs. (4.3) with (4.4), we get:

$$k_{eff} = \sigma\, e\left(T_1^4 - T_2^4\right)t/2(n+1)(T_1 - T_2)\ \text{kW/mK} \tag{4.5}$$

Inserting some figures, using $T_1$ = 293 K, e = 0.02 (as for aluminium), and (n + 1)/t = 3000 reflectors per metre, then:

$$k_{eff} = 1.4/(T_1 - T_2)\ \text{mW/mK} \tag{4.6}$$

For $T_2$ = 77 K

$$k_{eff} = 6.5\ \mu\text{W/mK} \tag{4.7}$$

This simple calculation shows that there is a theoretical lower limit to the effective conductivity of reflective insulants using aluminium, in the region of 6 μW/mK.

In theory, lower values should be attainable with gold or silver coated reflectors with e < 0.01 and with a packing density of reflective layers greater than 3000 per metre.

In practice, there are other contributions to the heat flow, via solid conduction, residual gas conduction and molecular desorption between the reflecting layers. The theoretical limit of 6 μW/mK therefore represents a target figure to aim for.

A number of different MLIs are now used commercially, as indicated in Table 4.3. By using spacer materials between the reflectors, either of fibre glass paper (with no filler) or nylon net, or by having dimples in aluminised plastic film, $k_{eff}$ values down to 30 μW/mK can be realised under test conditions, but somewhat higher values are realised in practice. Nevertheless, these practical values of $k_{eff}$ are at least 10 times lower than evacuated powders.

Table 4.3 gives typical quoted values of effective mean thermal conductivities between 300 and 77 K of some MLIs [9].

**Table 4.3** Effective thermal conductivities of multi-layer insulations between 300 and 77 K, in μW/mK

| | Layers (cm) | $k_{eff}$ (μW/mK) |
|---|---|---|
| 0.01 mm aluminium foil + Dexter paper | 20 | 52 |
| 0.01 mm aluminised Mylar + Dexter paper | 9 | 200 |
| 0.006 mm aluminium foil + nylon net | 11 | 34 |
| NRC2 crinkled aluminised Mylar film | 20 | 28 |
| 0.01 mm aluminium foil + carbon-loaded fibreglass paper (Southampton MLI) | 30 | 8–10 |
| 0.01 mm aluminised Mylar + carbon-loaded fibreglass paper | 30 | 26–36 |

By paying careful attention to reducing the gas pressure between the reflective sheets, a further 3-fold reduction in $k_{eff}$ values, down to below 10 μW/mK, can be realised. These lower $k_{eff}$ values may be obtained by the following steps:

(1) Reducing the residual gas pressure in the vacuum space containing the MLI to the minimum by extensive vacuum pumping, and by the strict avoidance of vacuum leaks.
(2) Avoiding molecular desorption within the MLI by high temperature, vacuum baking at 200–300 °C of all the MLI components before assembly or wrapping. Baking at 100 °C will remove adsorbed water, but not adsorbed gases like hydrogen or helium.
(3) Avoiding the use of plastic materials, which contain a great deal of adsorbed molecules and cannot be baked out at sufficiently high temperature to remove them.
(4) Using cryo-pumping materials within the MLI, for example self-pumping spacer material such as activated carbon-loaded fibre-glass paper, to mop up residual molecular desorption after steps (2) and (3) have been carried out. Fibreglass paper, without carbon loading, will also act as a self-pumping spacer after steps (2) and (3), but it has limited adsorption capacity.
(5) Using an additional getter of cryo-pumping materials, within the vacuum space, as a back-up to the high vacuum needed anyway in the depths of the MLI between the reflectors.

For example, when all five steps are taken together, $k_{eff}$ values down to 7 μW/mK have been regularly obtained at Southampton, using aluminium foil and carbon loaded fibre-glass paper spacer [10].

The main problem with present-day MLI is the highly anisotropic nature of its k-values. The k-value for aluminium in the plane of the insulation is in the region of 200 W/mK, some 2 million times larger than the $k_{eff}$ being aimed for normal to the plane. This makes application, to spherical and non-uniform axial geometries, a major problem in order to realise effective k-values below 100 μW/mK during manufacture. The anisotropy is reduced by using aluminised plastic film, with an aluminium thickness of less than 0.01 μm, as the reflector, but the effective k-value is then raised by its poor desorption properties. Some fundamental work is needed to reduce the anisotropy of aluminium foil reflector MLI and avoid the use of plastics with their poor degassing properties.

## 4.4 Conduction Down the Neck and Vapour Cooling

From 1954 onwards, research in cryogenics became increasingly focussed on developing the technology and infrastructure for liquid hydrogen and oxygen fuelled space rockets. A major contributor was the newly created NBS Cryogenic Engineering Laboratories, Boulder, Colorado, USA under the leadership of Russell Scott.

In 1959, Scott first published his book entitled "Cryogenic Engineering" [1] in which he was able to distill the thinking, findings and experiences gained from the first five years (13 years in the 1967 reprint) of unclassified operations of his laboratory colleagues. In his chapter on *Storing and Transporting Liquefied Gases*, he considered the problem of heat conduction down the neck of liquid hydrogen and helium vessels, and the concept of the vapour cooled vent tube. He observed that the cold available in a cryogenic vapour for absorbing heat inflows can be many times larger than the latent heat of evaporation of the cryogenic liquid. He further stated "in those cases in which heat conduction through the supports or fill and vent tube is a substantial part of the total, it may be worthwhile to use the escaping vapour to intercept part of the heat." And again, "the maximum possible saving from this process can be computed upon the assumption that the heat transfer between the vapour and the tube is perfect—that both have the same temperature at each level."

We now know that it is both worthwhile and easily possible to intercept **all** the conducted heat down the neck. We also know that the heat transfer between vapour and tube wall is enhanced 10-fold or more, as a result of the steep temperature gradient in the neck, and is close to the "perfect" assumption Scott made.

The first indication of this extraordinarily powerful heat transfer mechanism was deduced in 1968 by Lynam and Scurlock [11] from experiments measuring vertical and horizontal temperature profiles in the necks of liquid helium Dewars. The horizontal profiles showed that the temperature difference between the centre of a vapour column and the neck wall was less than 0.1 K at the 7.9 K level and no more than 0.3 K at the 57.5 K level. In other words, there was a qualitative indication of a natural horizontal convection process providing a strong heat transfer mechanism from the neck wall to all of the cold vapour column; much stronger than conventional theory suggested.

The vertical temperature profiles were measured for different evaporation rates of the liquid helium produced by an immersed heater, and therefore for different mass flows of cold vapour. These profiles showed what happened to the neck wall temperature in a dramatic way (see Fig. 4.4).

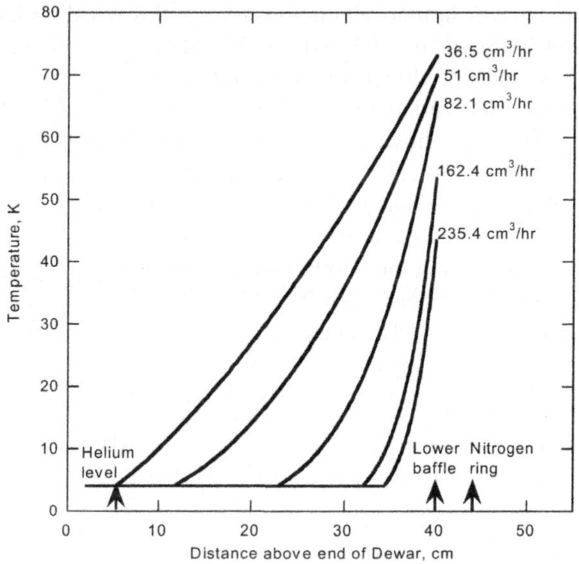


**Fig. 4.4** Variation of vertical temperature profiles with evaporation rate

As the vapour mass flow increased, so the temperature at any level on the neck wall decreased—as might be expected. However, at the lower end of the neck, a previously unexpected event was observed and can be seen in Fig. 4.4.

As the cold vapour mass flow increased, the temperature gradient at every level in the neck wall was also decreasing. In other words, the vertical profiles were showing that the increased vapour flow was decreasing the conducted heat flow down the neck wall.

In particular, the increased mass flow was decreasing the temperature gradient at the level of the liquid surface and hence the conducted heat flow into the liquid, $W_0$.

As the mass flow increased further, the temperature gradient at the liquid surface became zero, and the liquid temperature of 4.2 K began to rise up the neck wall.

The conducted heat flow into the liquid had indeed become zero, i.e. $W_0$, was identically zero.

The 4.2 K cold point continued to rise up the wall with increasing cold vapour mass flow. On reducing the vapour mass flow, the 4.2 K point moved down the wall towards the liquid surface. It therefore appeared that there was a particular vapour mass flow $M_0$ at which the 4.2 K cold point was just starting to lift off the liquid surface, and the neck conducted heat current was zero. In other words, this particular cold vapour mass flow $M_0$ was just sufficient to cool the neck and remove 100% of the conducted heat flow.

It follows that there appears to be a 1–1 correspondence between $M_0$ and $W_0 = 0$, a feature which can be incorporated into the design of cryogenic storage vessels to attain minimum boil-off.

The next step was to consider carrying out some calculations based on perfect heat transfer between vapour and neck wall, together with the above findings from the vertical temperature profiles.

Indeed, similar calculations for stainless-steel neck walls had already been carried out by Scott in 1963 and are described in his book. However, the diagrams Scott produced are difficult to interpret for the design of Dewars, since he plotted $W_0$ against the variable L/A, the neck form factor, for different vapour mass flows M.

They can, however, be turned round to aid Dewar or container design directly. The trick is to calculate and plot the vapour mass flow M against L/A for different values of the independent variable $W_0$. (L is the neck wall length, A is the cross-sectional area of the neck-wall) [12].

Transposing the Scott diagrams in this way, the performance diagram for helium Dewar necks between 4.2 and 77 K is obtained, as shown in Fig. 4.5a as a log-log plot of W (= λM) against L/A.

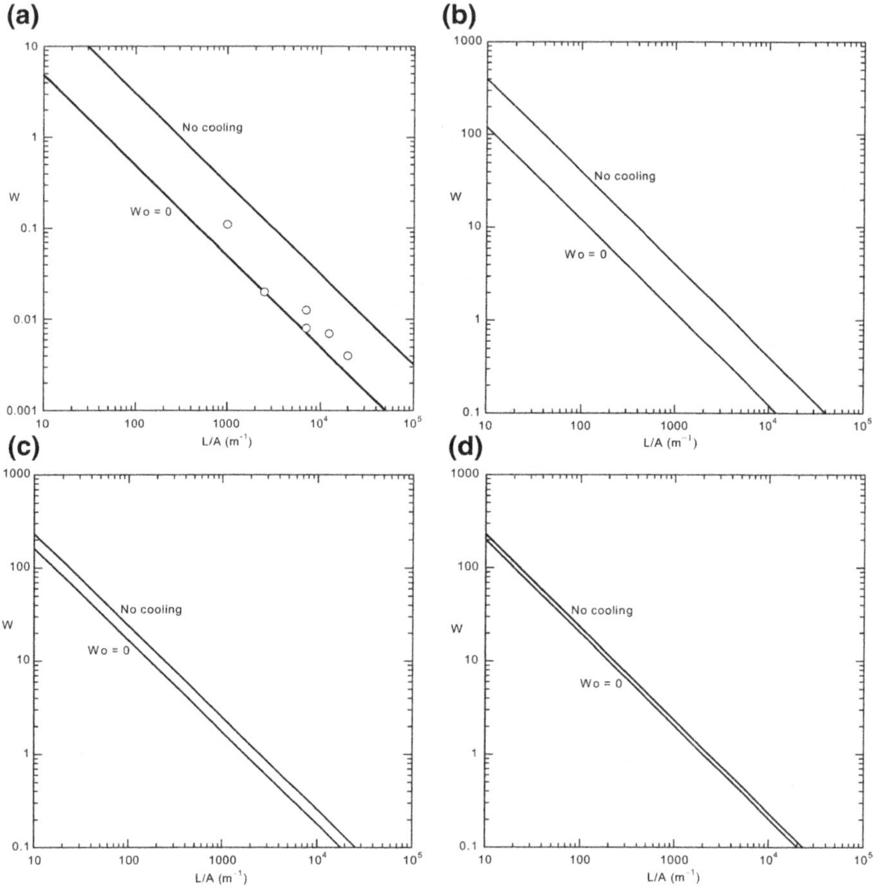


**Fig. 4.5** **a** Performance diagram for LHe vessels 4–77 K, O—performance of LIN shielded LHe Dewars. **b** Performance diagram for liquid hydrogen vessels 20–300 K. **c** Performance diagram for LIN vessels 77–300 K. **d** Performance diagram for liquid methane vessels 112–300 K

Similar diagrams can be easily produced for other cryogenic liquids, including hydrogen, nitrogen and methane, as shown in Fig. 4.5b, c and d respectively. The vertical axis represents the total heat flow W (= $\lambda M$) entering the liquid, or the mass flow M (= $W/\lambda$) of vapour passing up the neck (where $\lambda$ is the latent of vaporisation), or the evaporation rate. The horizontal axis is a plot of the stainless-steel neck form factor L/A.

In the diagram, there are two parallel lines with the slope of −1 which represent two limiting conditions as follows:

(a) When there is no heat transfer between vapour and neck wall—the worst case design scenario.
(b) When there is no conduction heat current $W_0$ reaching the liquid at the bottom of the neck. i.e. $W_0 = 0$. This is the target design condition when the vapour mass flow is just sufficient to provide 100% neck cooling. This vapour mass flow is about 10 times lower than with no neck cooling for the same neck geometry.

There is also a possible third parallel line when the only heat flow entering the liquid is the conduction heat flow down the neck, i.e. $W_0 = \lambda M_0$. This is the theoretical minimum possible evaporation rate $M_0$ attainable for a given neck form factor L/A, which is only about 30% lower than the (b) condition ($W_0 = 0$).

## 4.5 Optimum Design for Minimum Loss of Cryogenic Liquid in a Storage Container

In practice, it should be possible to design any low-loss storage container, vessel or Dewar to meet condition (b) with $W_0 = 0$.

Since radiation down the neck can be reduced to zero, or to a negligible quantity, by means of vapour cooled shields higher up in the neck, the optimum design criterion for a selected L/A is met by reducing the heat in-flow through the insulation space down to but not less than the value given by the $W_0 = 0$ line, or working line, on the performance diagram. If the insulation heat in-flow is reduced any further, then $W_0$ will become finite and very little further reduction in evaporation rate will be achieved.

A simple crosscheck that the optimum design criterion is being achieved by a cryogenic container is to measure the vertical temperature gradient in the vapour (and hence the gradient in the neck wall) at the bottom of the neck. For the $W_0 = 0$ condition to be met, this gradient should flatten out to zero at the base of the neck.

If the gradient is not zero, then the neck is too short and needs to be lengthened to meet the optimum design criterion.

Similar performance diagrams for other cryogenic liquids and neck-wall materials can be used, for designing necks and unwetted walls of storage vessels and tanks, in the same manner, when the ratio of available sensible heat of the cold vapour to the latent heat of vaporisation is greater than about 0.5 as can be seen in Table 3.1. The ratio is particularly high for liquid helium and hydrogen, but is also significantly greater than 0.5 for liquid neon, nitrogen, air, argon and oxygen as well as for liquid methane and LNG.

There are many commercial LIN vessels which do not meet this design criterion. Their loss rates are much too high because their necks are too short. Extending their necks by only a few centimetres is all that is required to reduce significantly, even halve, their loss rates.

Figure 4.5a also shows some performance points for six working liquid helium Dewars with LIN cooled shields in the insulation space surrounding the liquid helium.

It can be seen that two Dewars, with L/A of 30 and 75 $cm^{-1}$ respectively, had evaporation rates of 25 and 9 ml/h, corresponding to heat inflows of 20 and 7 mW respectively, with $W_0$ close to zero. On the other hand, the other four Dewars had about twice the evaporation rate for the zero $W_0$ condition.

Calculating the heat flow from the polished copper, LIN cooled shields suggested that only about one half of the observed heat inflow was coming from the LIN cooled shields. The source of this residual heat flow remained a mystery for many years and was not discovered until the late 1970s.

We now know that this residual heat flow arises from reverse convection in the core of the vapour column and determines the lowest possible loss rate which can be achieved in practice. See Sect. 4.6.3 below.

## 4.6 Convective Heat Flows into the Vapour and Liquid

### *4.6.1 Convective Circulations*

In general, density stratification, with hotter fluid, vapour or liquid, above colder fluid, acts so as to oppose vertical natural convection and vertical mixing between layers. In the absence of any heat flow, this convective stability applies to both vapour and liquid.

However, local heating changes the convection picture in a complex manner, which we are only just beginning to understand from experimental studies with mainly LIN, but also with LHe [13–16]. Convection in a cryogenic situation has many features common to convection in the atmosphere, and that is exceedingly complex as we all know.

Basically, in a closed system (see Fig. 4.6a for an example of the complex convection in nitrogen vapour above LIN), local heating generates an upward convective flow at the wall, which is additional to and possibly greater than the boil-off mass flow. At the same time, this upward flow induces downward flow in the vapour core, so that the induced downward mass flux is comparable to the additional upward mass flux.

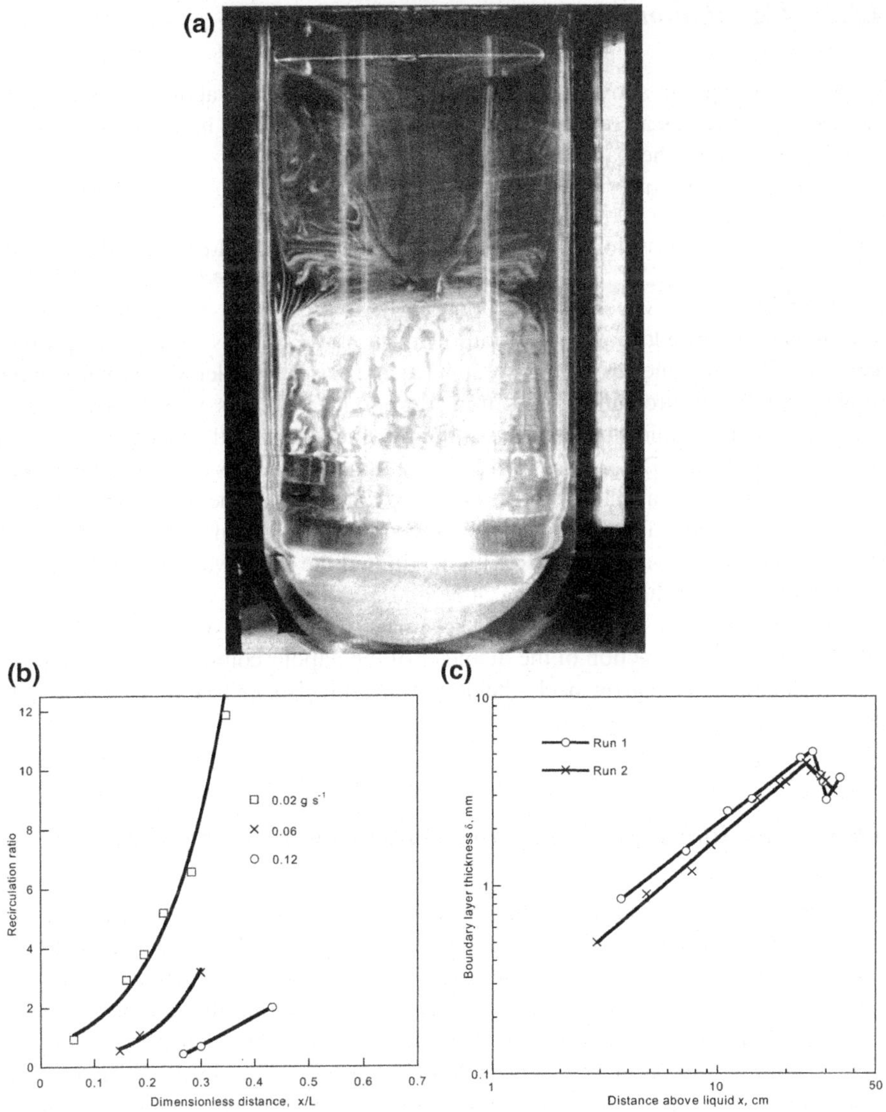


**Fig. 4.6** **a** Flow visualisation with smoke tracer, showing complex, double thermosyphon, vapour convection above LIN in 24 cm diameter Dewar. **b** Variation of recirculation ratio (negative mass flow/boil-off mass flow) with distance above liquid at different boil-off rates in 14 cm diameter column, L = 35 cm. **c** Variation of vapour boundary layer thickness with distance above liquid in 14 cm diameter column. In the lower region of the vapour column, δ is proportional to x

### 4.6.2 Circulation in the Vapour

In the vapour column above a boiling liquid, local heating of vapour in contact with the neck wall creates an upward boundary flow which in turn induces a downward flow in the core of the vapour.

From mass continuity, the following equation applies:

$$\text{Total downward mass flow} = \text{Total upward mass flow} - \text{Boil-off mass flow} \quad (4.8)$$

Alternatively, the vapour convection can be expressed as a recirculating mass flow R(vapour) in addition to the boil-off mass flow. Studies of velocity profiles across horizontal planes in vapour columns, using Laser Doppler Velocimetry, have shown that this recirculation mass flow R(vapour) increases with height from a finite value at the liquid surface to a maximum which can greatly exceed the boil-off (see Fig. 4.6b from LDV studies over LIN); and then decreases to zero at the roof of the vessel. Flow visualisation studies have revealed that elements of the core flow are sucked into the wall boundary layer flow at all levels down to the liquid/vapour interface; hence, the variation of R with height and the linear variation of boundary layer thickness with height (see Fig. 4.6c and Sect. 4.7).

Looking more closely, the downward mass flow of the recirculation is distributed over a large fraction of the diameter of the vapour column, with consequent small velocities. In a narrow neck, the downward flow is localised in a high velocity jet.

### 4.6.3 Residual Heat Flow from Downward Flowing Vapour

In all cases, the induced recirculating downward mass flow has been discovered to carry heat towards the liquid/vapour interface. This "residual" convective heat flow contributes to the boil-off, and, in the case of good insulation practice reducing all other heat flows, becomes the major source of heat entering the liquid. (The term "residual" is used to distinguish this heat flow from all other sources).

Unfortunately, the heat flow arising from the recirculating downward mass flow penetrating to the liquid surface is not very well understood and cannot therefore be quantified theoretically. Neither has any experimental correlation between R(vapour), vapour column diameter and associated heat flow ever been produced.

In principle, physically isolating the vapour core from the wall boundary layer flow would be expected to stop the suction of core vapour into the boundary layer. Such a flow isolator would need to create an annular space of some 10 mm wide, adjacent to the neck for the boundary layer flow, and have zero thermal conductance. In practice, such a device introduced into the neck of a Dewar vessel has never been observed to further reduce the liquid boil-off and is not recommended.

Attempts to divert the downward core flow into the upward boundary layer flow above the liquid surface have also never been successful. It was not realised that the vapour recirculation is driven by the heat absorbed from the neck wall by the boundary layer flow, and not by the boil-off mass flow.

Referring back to the performance points in Fig. 4.5a for the four liquid helium vessels, the magnitude of this residual heat flow was the same as that through the insulation space on to the whole of the wetted surface i.e. the total heat flow into the liquid is twice as large as the design figure without the residual heat flow.

In earlier work, Lynam and Scurlock [10] demonstrated how this residual heat flow can be separated experimentally from the heat flow through the insulation space. Using an immersed heater in the liquid helium, and measuring the evaporation rate as a function of heater power, they obtained the result shown in Fig. 4.7. Above a certain heater power, the graph is linear with a slope of 1 as might be expected. However, extrapolating the linear section back to the zero heater condition, shows a quantity $M_1$ corresponding to the heat leak through the insulation, which is independent of heater power as expected, together with a quantity $M_2$ which diminishes to zero with increasing heater power and therefore increasing mass flow up the neck. This quantity $M_2$ is believed to correspond to the residual convective heat flow produced by the reverse convection in the vapour column. The experimental results show that $M_1$ and $M_2$ are similar in magnitude, as mentioned above.

It would appear, therefore, that it is this residual convective heat flow, of the downward flowing vapour core, which determines the minimum boil-off rate attainable in practice.

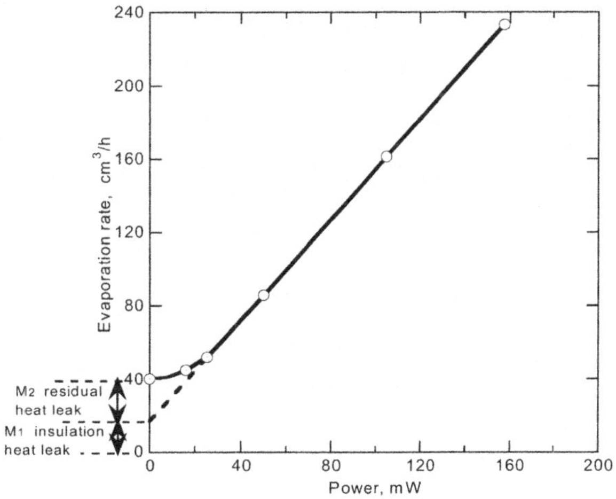


**Fig. 4.7** Helium evaporation rate as a function of the power dissipated in the submerged heater

### 4.6.4 Convective Circulation in the Liquid

In the liquid, the heat flow through the container wall is insufficient to produce local boiling. Instead, for depth/diameter ratios of unity or greater, the heat is absorbed by a convective boundary layer flow up the wall to the liquid surface. At the surface, the liquid flow, by this point superheated, turns over through a right angle and moves radially inwards when evaporation occurs as discussed in detail in Chap. 2. At the centre, the flow turns through another right angle forming a strong, downward moving, central jet. This jet is dispersed by mixing with the core (see Fig. 4.8a).

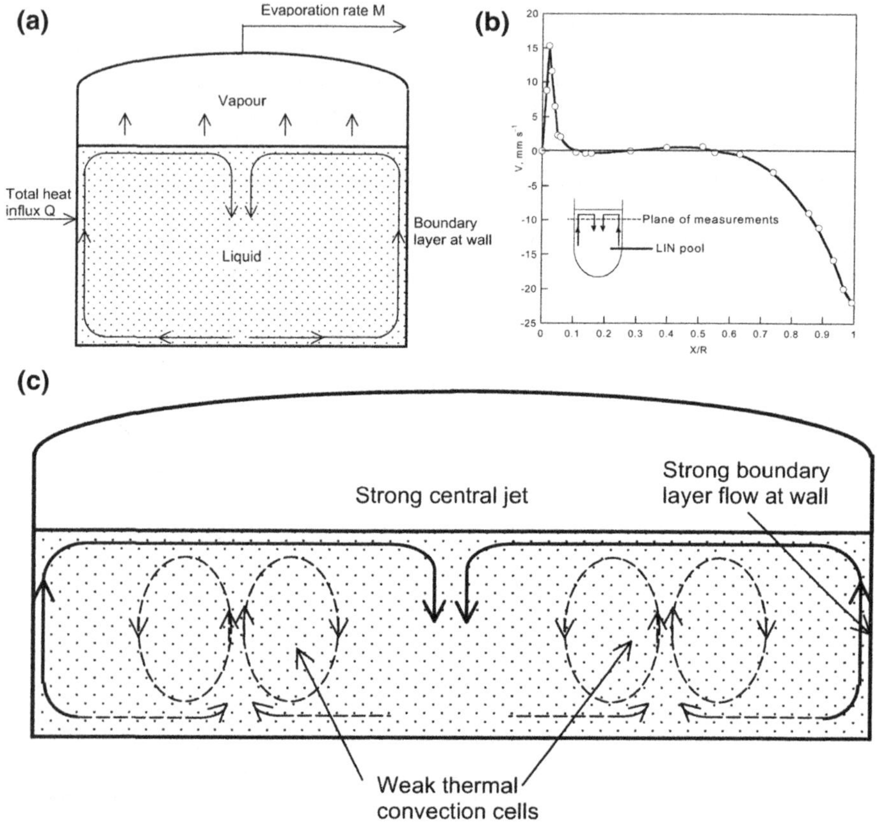


**Fig. 4.8** **a** Liquid circulation in storage tank via boundary layer at wall. Depth/diameter ratio greater than unity. **b** Vertical velocity profile by Laser Doppler Velocimetry in an evaporating liquid nitrogen pool. X = distance from wall, R = pool radius, depth 13 mm below surface. **c** Possible liquid convective flow in tank with small depth/diameter ratio, showing strong boundary layer flow and multiple thermal convection cells

Again, as in the vapour, the boundary layer flow up the wall induces a reverse, downward flow in the core of the liquid (see Fig. 4.8b) such that the following continuity of mass flow equation applies at the liquid surface:

$$\begin{aligned}\text{Total downward mass flow (core)} &= \text{Total upward mass flow (boundary layer flow)}\\ &\quad - \text{Surface evaporation mass flow (boil-off)}\end{aligned} \tag{4.9}$$

Alternatively, the convective flow can be seen again as a recirculating flow R (liquid) in addition to the heat-inflow generated mass flow. Both heat-inflow generated mass flow and R(liquid) increase with height from the bottom of the liquid, the former reaching a maximum at the surface, while R(liquid) probably reaches a maximum at some point below the liquid surface.

For depth/diameter ratios of 0.5 or less, the liquid convective flows induced by the heat-inflows are more complex, with additional convective thermals as well as the boundary layer flow at the wall carrying heat from the tank floor up to the surface. Each rising thermal is surrounded by a shell of sinking, colder liquid, the two flows making a convective cell such that a similar continuity of mass flow equation applies thus:

$$\begin{aligned}\text{Total downward mass flow (around the core)} &= \text{Total upward mass flow (core thermal)}\\ &\quad - \text{Surface evaporation mass flow}\end{aligned} \tag{4.10}$$

Since the diameter of the convective cells is approximately equal to the depth of liquid, then for a depth/diameter ratio of 0.5, it is possible to envisage up to six convective cells with six thermals in a hexagonal pattern carrying heat from the tank floor to the surface, in addition to the much stronger convective flows up the wall of the tank and down the central jet from the surface (see Fig. 4.8c).

## 4.7 Vapour Convection at the Unwetted Walls

At any level in the ullage (vapour) space, the wall is always hotter than the vapour and injects heat into the vapour. The resulting natural convection boundary layer flow up the wall cools the wall and absorbs heat being conducted down the wall into the liquid. If the neck is long enough, and/or the boil-off vapour mass flow is large enough, then all the conducted heat W in the wall is absorbed by the sensible heat of the vapour rising up the neck in the boundary layer flow.

The boundary layer flow is quite different to that normally met at ambient temperatures, where the thickness varies as

$$d \sim z^{0.25} \tag{4.11}$$

and grows very slowly with height z remaining in the region of 0.1–0.2 mm [17].

In the vertical temperature gradient pertaining to the cryogenic vapour column near the liquid surface, the overall convective flow is in the so-called “developed stable region”. The boundary layer thickness varies much more rapidly with height as

$$d \sim z \tag{4.12}$$

partly due to the rapid temperature and property (density and viscosity) variation with height and partly through the strong suction of the boundary layer flow drawing elements of vapour from across the whole of the column into the boundary layer.

This suction leads to a progressive increase in boundary layer mass flow and induces the convective recirculation, with accompanying reverse flow in the core of the vapour column. Consequently, the rising boundary layer grows linearly with height and becomes much thicker than ambient temperature boundary layers, with a thickness of the order of 10 mm (see Fig. 4.6c from studies on nitrogen vapour above LIN).

In the upper part of the ullage space, the convective flow becomes unstable with internal oscillations, and the boundary layer becomes progressively thinner. This region is called the “unstable stratified” region and is where the vapour recirculation decreases with increasing height [18].

## 4.8 Enhanced Convective Heat Transfer in Vertical Temperature Gradients

### *4.8.1 Use of Enhanced Convective Heat Transfer*

The success of the design criterion for vapour cooled necks depends on the heat transfer between vapour and neck wall being extremely good. Investigations at Southampton have demonstrated that the heat transfer coefficient, h, is indeed enhanced in a vertical temperature gradient, via a stratification factor S, in the empirical relation:

$$h = h_0(1 + S^n) \tag{4.13}$$

where $h_0$ is the coefficient in the absence of a vertical temperature gradient,

$$S = C(dT/dz)\Delta T \tag{4.14}$$

$$n = 5/12 \text{ for } S > 3. \tag{4.15}$$

C is a characteristic length, $\Delta T$ is the mean temperature difference between vapour and neck wall, and (dT/dz) is the local vertical temperature gradient [19–22].

At ambient temperatures, vertical temperature gradients are relatively small and associated heat transfer enhancement is small. However, in the neck of a helium Dewar, S can easily be as high as 200, leading to a tenfold enhancement of the local heat transfer.

For example, in a typical neck, $(dT/dx) = 1000$ K/m, $C = 0.05$ m, and $\Delta T = 0.5$ K, giving $S = 200$, $h = 10.2\ h_0$ and the rate of heat transfer is enhanced 10 fold.

The fluid dynamical modelling of the enhanced heat transfer to the neck wall by the thick boundary layer flow presents one of several challenges pertaining to the observed complex convective behaviour of a cryogenic vapour column.

### *4.8.2 Enhanced Cooling of Current Leads to Superconducting Magnets*

The 10 fold heat transfer enhancement can, of course, be applied advantageously to the design of current leads to superconducting magnets whereby the bare copper leads are positioned close to the neck wall and cooled by the boundary layer flow. Figure 4.9 shows a successful design for a pair of 500 A leads in a spiral length of 15 cm between liquid helium and ambient temperatures.

**Fig. 4.9** High thermal efficiency 500 A spiral superconducting magnet current leads. The total length is 15 cm between 300 and 4 K

The enhancement has also been employed in the design of the 15,000 A current leads at CERN, enabling an enormous reduction in refrigeration power for the LHC ring.

### 4.8.3 *Cryocooler/Condensers with Distributed Cooling*

Cryocoolers have generally been constructed to provide spot cooling at two or three temperatures in stages. The thermodynamic efficiency of cryocoolers, and as cryocondensers for zero net boil-off, is therefore very poor because of the large $\Delta T$ in the heat transfer processes by the spot coolers.

The introduction of distributed cooling can reduce the large $\Delta T$, and increase the thermodynamic efficiency significantly. We know that the vapour cooling of neck walls is very efficient via distributed cooling of boundary layer flow at the neck wall. This cooling is enhanced by the use of a set of horizontal fins which, in addition, absorb the radiation element of the heat flow down the neck, again by distributed cooling (see Fig. 4.10).

Thus, applying this understanding to cryocoolers, by adding a set of horizontal fins, revealed how the refrigeration/condensing performance could be substantially improved; i.e. almost doubled at the first attempt [23]. Further studies are needed to make even better use of distributed cooling for cryocoolers/condensers.

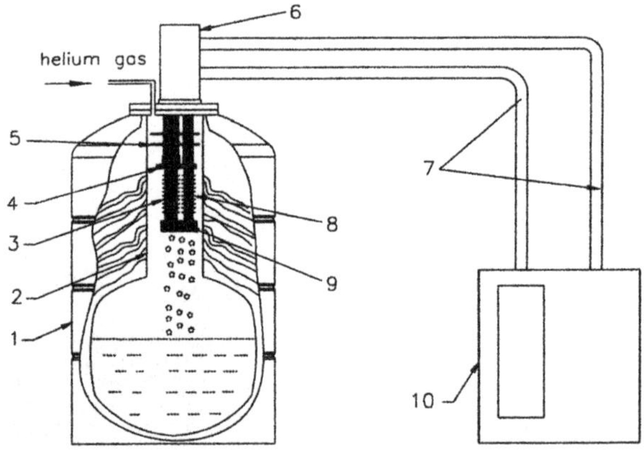


**Fig. 4.10** Prototype cryocooler with fins (1) helium Dewar, (2) Dewar neck, (3) fins on the 2nd stage regenerator, (4) 1st stage cooling station, (5) radiation shields/fins on the 1st stage regenerator, (6) pulse tube cold head, (7) flexible lines, (8) fins on the 2nd stage pulse tube, (9) 2nd stage cooling/condenser and (10) compressor

## 4.9 Multi-shielding: The Use of Multiple Vapour Cooled Shields in the Insulation

### *4.9.1 Converting 'A' Heat-Inflows to 'B' Heat Inflows*

We have already seen in Chap. 3 how the cooling power of the cold vapour as it warms up to ambient temperature greatly exceeds the latent heat of vaporisation, by 75 fold for liquid helium, 8 fold for liquid hydrogen and 3 fold for liquid neon.

The vapour cooling power (i.e. the enthalpy gain or sensible heat between boiling point and ambient) is the same magnitude as the latent heat of vaporisation in the cases of LIN, LOX and LNG.

Thus, the magnitude of 'B' heat in-flows which can be absorbed by the vapour flow can be significantly greater than, or at least comparable with, the 'A' heat in-flows absorbed by the liquid evaporating and generating the same vapour flow.

This fact is particularly important in the design of low-loss vessels for liquid helium and liquid hydrogen. It can also be significant in the advantageous design of minimum loss vessels for LIN, LOX, LA, LNG and liquid air. For the low-loss storage of liquid ethane, ethylene and higher hydrocarbons, the available vapour cooling for absorbing 'B' heat in-flows is reduced; the design for minimum boil-off needs to match this reduction by reducing the 'B' heat inflows by, for example, employing a longer neck or unwetted tank wall.

In general, however, if the vapour cooling power is greater than 50% of the latent heat, the cold vapour flow has a cooling capacity in excess of that required for cooling radiation shields in the neck, as well as cooling the neck wall to achieve the $W_0 = 0$ condition. This excess cooling capacity can be used most advantageously to absorb some of the 'A' heat in-flow through the insulation space, via conversion into 'B' heat in-flow, in one of the following ways (as in Fig. 4.10):

(a) Replacing the LIN cooled shield enclosing the LHe or LH2 inner vessel with a single shield, vapour cooled to about 77 K.
(b) Using a second vapour cooled shield at a higher temperature, say about 120 K or 150 K.
(c) Using more than two vapour cooled shields, between three and a maximum of ten in practice, distributed through the insulation space so as to make maximum use of the vapour cooling capacity, all the way from the liquid boiling point up to ambient temperature.

The term "multi-shielding" is generally employed to describe the use of two or more vapour cooled shields, as in (b) and (c) in Fig. 4.11 [24, 25].

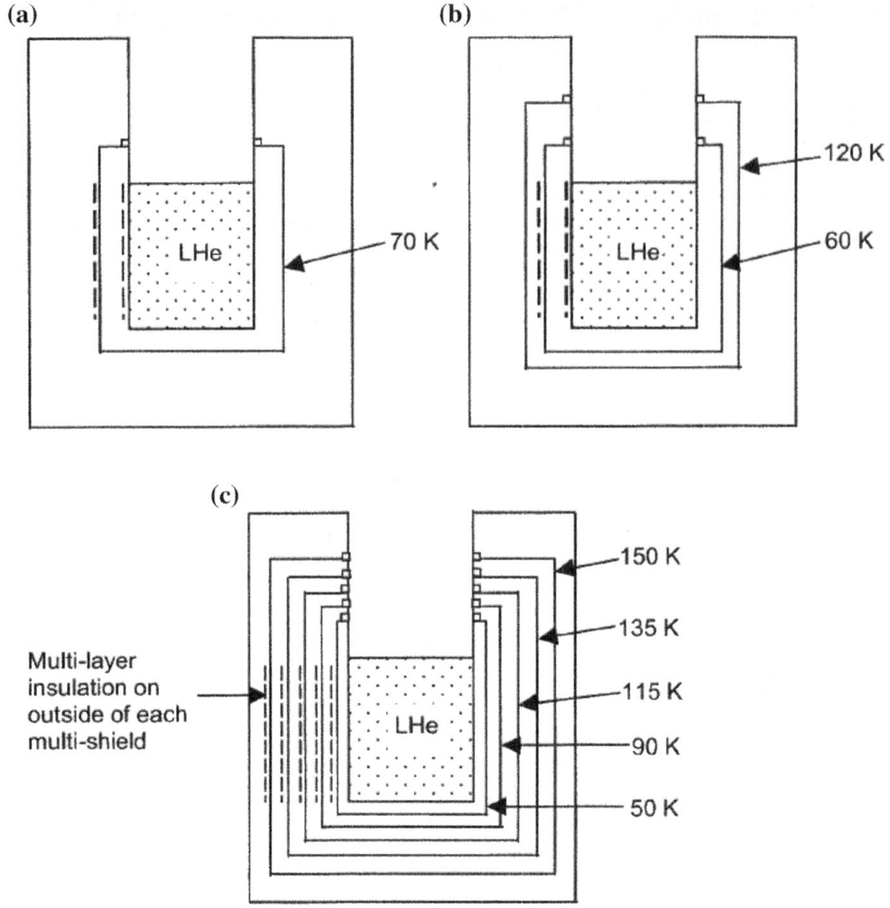


**Fig. 4.11** **a** Single vapour cooled shield. **b** Two vapour cooled shields. **c** 5–10 shields (multi-shielding)

### 4.9.2 Enhanced Heat Transfer at Thermal Contact Rings

Vapour cooling of the multi-shields is easily achieved because there is very good convective heat transfer between neck wall and the boundary layer flow of cold vapour immediately adjacent to the wall. Additionally, the vertical temperature gradient in the vapour further enhances the convective heat transfer, as described in Sect. 4.8 above, via two sources (a) the local temperature gradient in the neck arising from heat conduction down the neck, and (b) the additional temperature gradient in the vapour arising from the heat in-flow through the thermal contact ring of the shield.

As a result, the contact ring need only be 10–20 mm in vertical extent around the circumference of the neck.

### 4.9.3 No LIN Shielding for LHe Systems

A major advantage of multi-shielding for LHe systems is that it does away with the requirement for a LIN cooled shield and associated LIN servicing. This can be achieved with little, if any, increase in LHe evaporation rate and is therefore very attractive to users of LHe when LIN supplies are expensive and difficult to manage.

### 4.9.4 Assembly of Multi-shields in Insulation Space

A perceived disadvantage lies with the more complex assembly of multi-shields in the insulation space, whether (a) with foam or glass-fibre, plus purge gas, on a large scale, or (b) with evacuated powder on a medium scale, or (c) with MLI on a smaller scale.

However, the use of lightweight, high thermal conductivity, aluminium foil for the multi-shields simplifies assembly since the insulation material, whether in the form of sheets or blankets, or as MLI, can be employed to provide mechanical support and thermal separation of the multi-shields. The shields also do not have to contain a vacuum or sustain any pressure difference whatsoever, thereby simplifying assembly even further.

As a result, multi-shielding is widely used in LHe and LH2 vessels with great success in terms of low-loss performance, reliability and cost.

### 4.9.5 Vapour Cooled Shields for LIN, LOX, and LNG Vessels

For LIN, LOX, liquid air and LNG vessels, the use of a complex assembly in the insulation space to achieve a reduction in boil-off rate may be marginal, on the basis of the increased cost. However, in the smaller, vacuum insulated vessels using MLI, the inclusion of one, or perhaps two, vapour cooled shields within the insulation may be fully justified, for example, in order to minimise boil-off when LNG is used as a portable source of fuel in road vehicles.

## 4.10 Other Sources of Heat into the Liquid

### 4.10.1 Radiation Funnelling

Ambient temperature radiation can funnel down neck tubes and pipelines by internal specular reflection, without significant diminution, directly into liquid baths. Even if the tubes are vapour cooled, the radiation is not absorbed at the walls of the tubes, during internal reflection. To reduce radiation funnelling, the inner surfaces must therefore be rough so as to promote diffuse reflection at the relevant infra-red wavelengths. It is also advisable to use radiation baffles and traps, in all neck tubes and lines entering a cryogenic system.

### 4.10.2 Low Conductivity Neck Tube Materials

The discussion of vapour cooled neck tubes, from Scott's theory and its transposition into the performance diagrams of Fig. 4.5, is based on the use of stainless steel as the low thermal conductivity neck-wall material. There are, of course, other suitable low k-value materials. Historically, the first material to be used was glass, either high borosilicate Pyrex glasses for general application, or low borosilicate Monax glasses for use in contact with helium gas.

Plastic composites are very attractive because of their strength and low k-values, but they tend to be porous causing degradation of the vacuum systems required for insulating small vessels.

### 4.10.3 Thermo-Acoustic Oscillations

The prevention of thermo-acoustic oscillations, or Taconis oscillations, after Taconis, Leiden University, The Netherlands, first studied these sources of heat into liquid helium in 1949 [26], is important because they can generate very large 'A' heat in-flows. It is now widely known that tubes or pipes with large temperature differences along them may contain thermally sustained acoustic oscillations, particularly, but not exclusively, related to liquid hydrogen and helium systems. These vapour phase oscillations are not adiabatic, or isentropic, because of irreversible heat flows to and from the containing wall, and a net heat current results along the tube towards the cold end.

The standard example is the liquid helium dipstick used for determining liquid levels in many cryogenic laboratories. This consists of a meter length of 3 mm thin-walled stainless steel tube, open ended at the cold end, and terminating at ambient temperature in a small, 15 mm diameter, bell-shaped fitting with a rubber diaphragm across the end. When the cold end is immersed in liquid helium, the oscillation frequency is about 2–4 Hz. When the cold end is raised out of the liquid, the oscillation frequency increases abruptly to about 10–20 Hz. With a finger pressed lightly on the diaphragm, the transition can be sensed very easily with practice and the helium liquid level measured to within 1 mm.

Thermo-acoustic oscillations may be accompanied by a low frequency hum, and can be stopped by considering them as an acoustic problem. The solution is a change in geometry at the ambient temperature end and/or the introduction of an acoustic impedance, such as a Helmholtz resonator, to damp out the oscillation.

It is interesting to note that the Space Shuttle rocket motor incorporates a cold oscillation damper, called a POGO suppressor, in the LOX pumping line between low pressure and high pressure turbo-pumps [27, 28].

Today, thermo-acoustic oscillations are being used to drive a new class of refrigerator, with no moving parts. This development requires an improvement in understanding wall heat transfer to oscillating vapour columns, which will in turn no doubt help to prevent unwanted oscillations in low-loss cryogenic systems. Thermo-acoustic refrigerators offer a low cost alternative to both pulse-tube and vapour compression, closed cycle refrigerators down to about 100 K.

### *4.10.4 Mechanical Vibrations*

One source of heat flow to mention is that due to induced mechanical vibration. Whether the vibrations are resonant or non-resonant, they are dissipative so that mechanical energy is turned into heat, creating an internal source of heat inflow into the liquid, which may be difficult to identify. The answer is to insulate the cryogenic system from the external sources of vibration, such as vacuum pumps and rotating or reciprocating machinery, by using flexible mountings and couplings.

### *4.10.5 Eddy Current Heating*

Another source of heating can arise from electromagnetic fields causing eddy-current heating in low resistivity materials such as copper, aluminium and superconductors. The answer is to:

(a) divide the materials so that eddy-current loops are minimised,
(b) use high impedance materials, such as plastic composites, where eddy-current heating cannot be induced,
(c) use low impedance electromagnetic screens at ambient temperature surrounding the cryogenic system,
(d) employ combinations of all three.

## 4.11 Summary of Insulation Techniques

1. Identify all heat-inflows and then reduce them to the same magnitude using insulation suitable for scale of operation.
2. Design system to convert 'A' heat in-flows into 'B' in-flows by using vapour cooled baffles, vapour cooling of necks and pipework entering liquid, and/or vapour cooled multi-shields.
3. Use radiation shields instead of plastic foam neck plugs.
4. Note minimum heat-inflow to liquid may be determined by reverse convection in the core of vapour columns.
5. For large, flat-bottom tanks, requiring under-floor heating to prevent frost heave, build in extra thickness of load-bearing insulation below the tank-floor.
6. Introduce distributed cooling in cryocoolers for improved refrigeration.

## References


1. Scott, R.B.: Cryogenic Engineering. Van Nostrand (1959), 6th reprint (1967)
2. Lynam, P., Proctor, W., Scurlock, R.G.: Reduction of the evaporation rate of liquid helium in wide necked Dewars. In: Bulletin of IIR, Commission 1, Grenoble, Annex 1965-2, p. 351 (1965)
3. Lynam, P., Mustafa, A.M., Proctor, W., Scurlock, R.G.: Reduction of the heat flux into liquid helium in wide necked metal Dewars. Cryogenics **9**, 242 (1969)
4. Boardman, J., Lynam, P., Scurlock, R.G.: Reduction of evaporation rate of cryogenic liquids using floating, hollow, polypropylene balls. Proc. ICEC3 Cryogenics **10**, 133 (1970)
5. Chun, S.: Application of the anti-sloshing blanket to LNG carriers and FLNG. In: 2012 AIChe, Spring National Meeting, Houston (2012)
6. Dewar, J.: Proc. Roy. Inst. **15**, 815 (1898)
7. Peterson, P.: The heat-tight vessel. Ph.D. thesis, University of Lund, Sweden (1951)
8. Hunter, B.J., Kropschot, R.H., Schrodt, J.E., Fulk, M.M.: Metal additives in evacuated powder insulations. Adv. Cryog. Eng. **5**, 146 (1959)
9. Kropschot, R.H., Schrodt, J.E., Fulk, M.M., Hunter, P.J.: Multilayer insulations. Adv. Cryog. Eng. **5**, 189 (1959)
10. Scurlock, R.G., Saull, B.: Development of multilayer insulations with thermal conductivities below 0.1 μW/cmK. Cryogenics **16**, 303 (1976)
11. Lynam, P., Scurlock, R.G.: Vertical and horizontal temperature profiles in the necks of helium Dewars. In: Proceedings of the Second International Cryogenic Engineering Conference, Brighton, p. 141 (1968)
12. Scurlock, R.G.: The design of low-loss liquid helium Dewars. In: Proceedings of the Second International Cryogenic Engineering Conference, Brighton, p. 144 (1968)
13. Boardman, J., Lynam, P., Scurlock, R.G.: Solid/vapour heat transfer in helium at low temperatures. In: Proceedings of the Fourth International Cryogenic Engineering Conference, Eindhoven, p. 310 (1972)
14. Boardman, J., Lynam, P., Scurlock, R.G.: Complex flow in vapour columns overboiling cryogenic liquids. Cryogenics **13**, 520 (1973)
15. Islam, M.S., Scurlock, R.G.: Qualitative details of the complex flow in cryogenic vapour columns. Cryogenics **17**, 655 (1977)
16. Beresford, G.: LDV in cryogenic vapour columns. PhD thesis, Southampton University (1983)
17. McAdams, W.H.: Heat Transmission. McGraw Hill (1954)
18. Boardman, J.: Heat transfer in vapour columns. PhD thesis, Southampton University (1974)
19. Chen, C.C., Eichhorn, R.: Natural convection from a vertical surface to a thermally stratified fluid. J. Heat Transfer **98**, 446 (1976)
20. Tritton, D.J.: Physical Fluid Dynamics. Van Nostrand Reinhold (1979)
21. Islam, M.S., Richards, D.J., Scurlock, R.G.: The influence of thermal stratification and flow interaction on the enhanced natural convection heat transfer at low temperatures. Cryogenics **19**, 319 (1979)
22. Islam, M.S., Scurlock, R.G.: Analysis of solid vapour heat transfer in helium vapour columns at low temperatures. Cryogenics **18**, 323 (1978)
23. Wang, C., Scurlock, R.G.: Improvement in performance of cryocoolers as condensers. Cryogenics **48**, 169–171 (2008)
24. Bejan, A.: Discrete cooling of low heat leak supports to 4.2 K. Cryogenics **15**, 290 (1975)
25. Li, Q., Eyssa, Y.M., McIntosh, G.E.: Discrete cooling of supports and MLI in helium Dewars. Adv. Cryog. Eng. **29**, 785 (1983)
26. Taconis, K.W., Beenakker, J.J.M., Nier, A.O.C., Aldrich, L.T.: Physica **15**, 733 (1949)
27. Rott, N.: Thermoacoustics. Adv. Appl. Mech. **E20**, 135 (1980)
28. Tward, E., Mason, P.V.: Damping of thermoacoustic oscillators. Adv. Cryog. Eng. **27**, 807 (1982)

# Chapter 5
# Multi-component Liquids

## 5.1 Differences Between Single-Component and Multi-component Liquids

Previous chapters have all been concerned with single component liquids. We have seen how non-equilibrium states of superheating in the liquid are necessary for the surface evaporation mechanism and how they have a significant bearing on evaporation instabilities. Furthermore, the superheating can lead to density stratification with a layer of hotter, less dense liquid sitting on colder, more dense liquid in a stable fashion.

In real life, most cryogenic liquids are multi-component mixtures and can be expected to have more complicated evaporation behaviours than a single component liquid. Examples of cryogenic liquid mixtures include:

- Liquefied air, which contains oxygen, nitrogen, argon, and traces of water, carbon dioxide and hydrocarbons,
- Liquid natural gases, LNG, which contain nitrogen, methane, ethane, and smaller quantities of carbon dioxide and higher hydrocarbons,
- Liquefied petroleum gases, LPG, which contain propane, butane and other hydrocarbons,
- Industrially produced mixtures, such as synthetic natural gases, refinery off-gases, and ammonia cycle off-gases,
- Pure liquids, such as nitrogen, argon and oxygen, which all contain small but finite traces of dissolved impurities such as carbon dioxide and hydrocarbons. On exposure to atmospheric air, liquid nitrogen will condense air and pick up oxygen and carbon dioxide in solution, and water as a particulate.

It is therefore important, when storing and handling cryogenic liquids, to understand the different evaporation behaviour of mixtures compared with that of pure liquids, and how this can affect the build-up and release of thermal overfill. Thermodynamically, the equilibrium state of a single component fluid is defined by

 
T. D. Bostock and R. G. Scurlock, *Low-Loss Storage and Handling of Cryogenic Liquids*, International Cryogenics Monograph Series,
https://doi.org/10.1007/978-3-030-10641-6_5

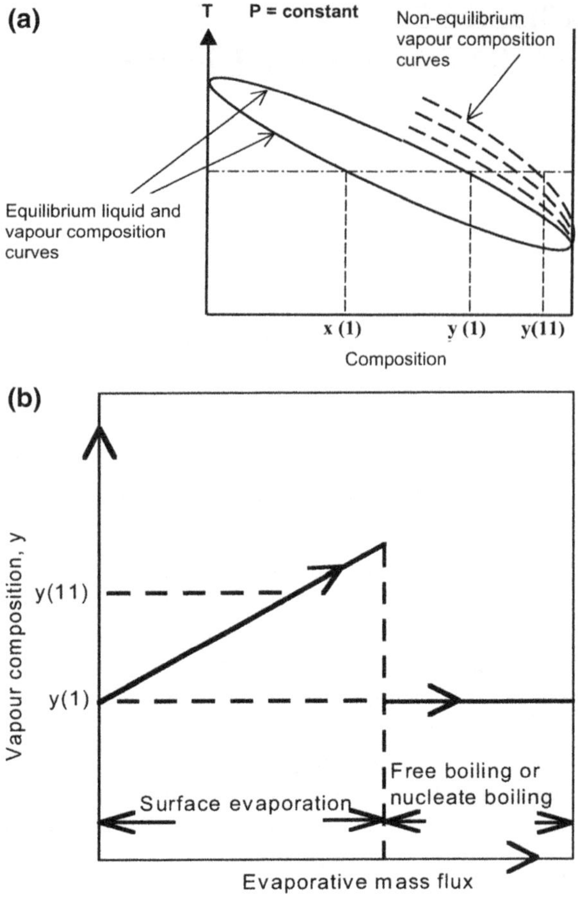


**Fig. 5.1** **a** Typical vapour and liquid (T, x) curves during equilibrium (free boiling) and non-equilibrium surface evaporation. **b** Deviation of vapour composition y(11) from free boiling value y(1) with increasing evaporative mass flux of liquid mixture

two out of the three variable parameters, pressure P, density ρ, and temperature T, the third parameter being defined by an equation of state.

With a multi-component fluid, an equilibrium state is defined by three or more parameters respectively out of four or more, including P, ρ, T, and x(1), x(2), etc. (where x(1), x(2), etc. are the molar fractions of component (1), (2), etc. in the fluid). A further complication is that vapour and liquid phases in equilibrium have different compositions.

This is illustrated in Fig. 5.1a by the typical equilibrium (T-x) diagram for a 2-component or binary mixture at constant pressure P, where equilibrium is achieved by a freely boiling liquid in contact with its boil-off vapour. "Freely boiling" is the boiling produced by a submerged or wall heater with a sufficiently large heat flux to generate nucleate boiling and multiple streams of bubbles rising through the liquid to the surface.

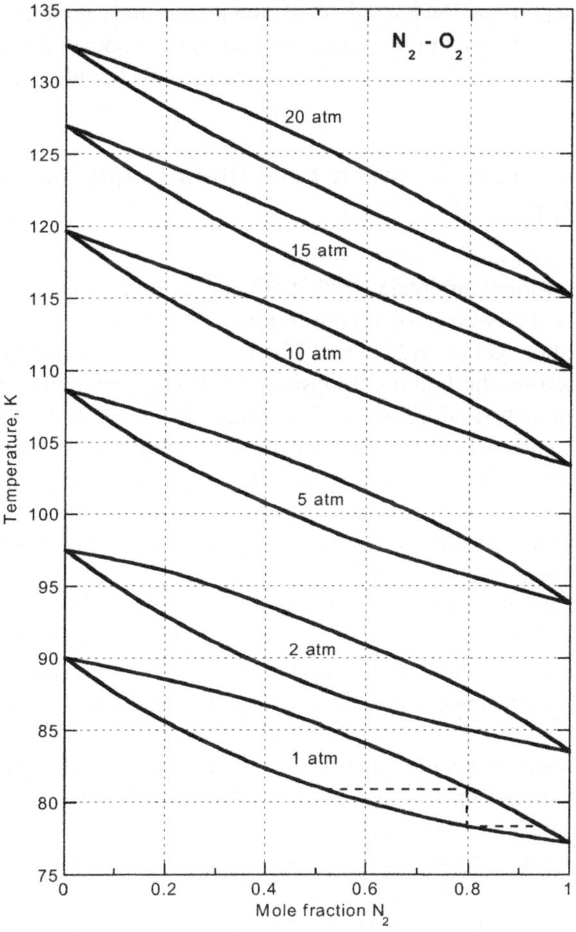


**Fig. 5.2** Temperature-composition diagram for oxygen-nitrogen mixtures

At constant pressure, the equilibrium concentration y(1) in the vapour of the lower boiling point component (1), which is in contact with a freely boiling binary mixture, is higher than its composition x(1) in the liquid.

This difference between vapour and liquid compositions is widely used in fractional distillation for separating a mixture into a lower boiling-point component and a higher boiling-point component.

Figure 5.2 shows the particular (T-x) diagram for oxygen-nitrogen mixtures at one atmosphere pressure, in which the wide difference in composition of vapour and liquid is demonstrated. It can be seen that the vapour in equilibrium with liquid air, (79% nitrogen 21% oxygen), contains less than 7% oxygen, and therefore presents a serious hazard from asphyxia as a breathing gas. Likewise, the condensation of air starts at 81 K and produces a liquid containing up to 48% oxygen, which presents a combustion hazard. Thus, the increasingly general use of liquid air as a refrigerant involves additional hazards over those associated with using LIN as a refrigerant (see Sect. 8.7.1 on safety procedures with liquid air).

## 5.2 The Difference Between Free-Boiling and Surface Evaporation (T-x) Data

It may not be widely appreciated that, under storage conditions with surface evaporation only, the well-known equilibrium (T-x) diagrams may not apply. The vapour composition y(11), y(22), etc. will not be the same as y(1), y(2), etc. as defined by measuring the liquid composition x(1), x(2), etc. together with T and P, and using the published equilibrium (T-x) data, (again, see Fig. 5.1a).

The difference arises because, unlike nucleate boiling with vapour bubbles rising rapidly through the bulk liquid and coming into equilibrium according to the (T-x) data, the surface evaporation of each molecular species is differentially controlled by its own diffusion mechanisms, across the surface sub-layer between the bulk liquid and the liquid-vapour interface.

Evaporation-limiting molecular diffusion rates will be determined in a complex manner by the double diffusion of each molecular species under both local thermal and concentration gradients; within region 1 of the surface sub-layer largely by concentration gradients, and within region 2 by the enormous temperature gradients of 5–10,000 K/m.

Because different molecular species in a mixture have different diffusion rates (inversely proportional to their respective molecular weights), as well as different evaporation coefficients, the relative concentration of species in region 1, the molecular evaporation region, will deviate from the bulk liquid with an expected relative increase in the lower molecular weight component (generally the lower boiling point component). In contrast, with nucleate or free boiling conditions, the local concentration of molecular species in the bulk liquid through which the bubbles are rising is the controlling mechanism for the vapour composition, rather than the surface sub-layer.

The difference in concentration of each molecular species in the surface sublayer will depend on the product of its own thermal diffusion coefficient and the temperature gradient across region 2, together with the product of its own self diffusion coefficient and the concentration gradient across the same thermal conduction layer. The heat flow through this layer can also be expected to be a factor in determining the temperature gradient and thickness of this layer.

Consequently, when the heat flows and associated evaporative mass flows increase, the concentration difference between interface and bulk liquid will increase. As a result, the vapour composition y(11) will deviate progressively from the equilibrium value y(1) as the evaporation rate increases, and will almost certainly contain more of the lower boiling point component (as indicated in Fig. 5.1b).

This change in vapour composition with surface evaporation rate has been studied in experiments at the Institute of Cryogenics, Southampton, after the effect had first been observed.

The consequences of non-equilibrium (T-x) values being different to published equilibrium values have significance in many areas. For example:

- When designing distillation columns with separation trays, or structured packing, where surface evaporation contributes to the fractional distillation.
- When using the composition of the liquid from a liquid mixture as the control variable for the vapour composition, such as trying, unwisely, to maintain operationally the composition of Breathing Air using a liquid oxygen-nitrogen mixture. The vapour will tend to be much richer in nitrogen than the equilibrium (T-x) value, thereby creating an unexpected asphyxiation hazard.
- When comparing published experimental vapour-liquid data from different authors. For example, liquid air data can show variations in compositions by up to 30% depending on the method of measurement.
- When using LIN in contact with atmospheric air. The air will condense, the nitrogen will evaporate preferentially and the oxygen concentration in the LIN will rise quite quickly.

## 5.3 Stratification in Cryogenic Liquid Mixtures

From Chap. 4, it can be seen how liquid convective circulation can lead to density stratification in single component liquids. In the absence of any continuous mechanical mixing of the single component liquid, the intrinsic mechanisms of superheated wall boundary layer flow, incomplete evaporation at the surface, and central downward jet of less superheated liquid, can lead to stratification into two layers. Each layer will be at a more or less uniform but different temperature, with a hotter, less dense layer on top of a colder, more dense layer.

As a guide to the associated temperature and density differences, see Table 5.1 which contains estimates of the changes of density with temperature and pressure along the saturation $P - T$ line for a whole range of cryogens. However, note that the density changes are for guidance only, since they relate to the equilibrium saturation condition and not to the superheated liquid from which evaporation takes place or to isenthalpic changes of liquid state.

Since the temperature difference across the liquid-liquid interface between the two layers is small, of the order of 0.1–1.0 K, the mixing effect of thermally driven molecular diffusion in the absence of any convective motion is relatively small. The associated density difference across the liquid-liquid interface acts so as to suppress local convective mixing and the stratification is therefore extremely stable.

However, the heating of the top layer by the wall boundary layer flow continues, leading to a continuing rise in temperature, or an increase in thickness accompanied by a downward migration of the interface, or a combination of the two.

**Table 5.1** Derivative properties of saturated liquid cryogens at NBPs

| | NBP (K) | ρ (kg/m$^3$) | $-(d\rho/dT)_{sat}/\rho$ (%/K) | $+(dT/dP)_{sat}$ (K/bar) | $-(d\rho/dP)_{sat}/\rho$ (%/bar) |
|---|---|---|---|---|---|
| Helium | 4.22 | 125 | 21 | 1.05 | 22 |
| Hydrogen | 20.28 | 70.8 | 1.6 | 4.24 | 6.8 |
| Neon | 24.55 | 1205 | 1.4 | 4.22 | 5.9 |
| Nitrogen | 77.31 | 806.8 | 0.57 | 10.8 | 6.1 |
| Argon | 83.80 | 1394 | 0.47 | 7.1 | 3.3 |
| Oxygen | 90.19 | 1141 | 0.43 | 12.26 | 5.2 |
| Methane | 111.67 | 422.4 | 0.40 | 14.3 | 5.7 |
| Ethane | 184.55 | 488.5 | 0.20 | 18.1 | 3.6 |
| Propane | 231.1 | 581 | 0.19 | 29.2 | 5.55 |

When the density difference between the two layers becomes large enough, of the order of 1.0% (corresponding to an approximate $\Delta T$ of 0.05 K for LHe, 0.6 K for LH2, 1.8 K for LIN, 2.5 K for LCH4, etc. from Table 5.1), the superheated wall boundary layer flow in the denser, lower layer suddenly has insufficient buoyancy and inertia to penetrate the liquid-liquid interface. The boundary layer flow is trapped in the lower layer and no evaporation can take place to release its superheat. Instead, the wall boundary layer turns over at the interface and its motion and superheat is locked into the lower liquid causing it to heat up instead. When this happens, the evaporation rate will fall as the first indication of a stratification effect and associated increase in thermal overfill via the unstable superheated state. As described in Sect. 2.2.3, this type of thermal overfill in a single component liquid may be released by a violent QHN boiling of the lower layer, with consequences such as the ejection of vapour mixed with liquid through the vents, and possible mechanical damage to the storage vessel.

Figure 5.3 illustrates how density stratification leads to the lower layer boundary layer flow failing to penetrate the liquid-liquid interface.

With a multi-component cryogenic liquid, the triggering and build-up of stratification can occur in a variety of ways, depending on the identity of the liquid components, and the previous history of the liquid elements.

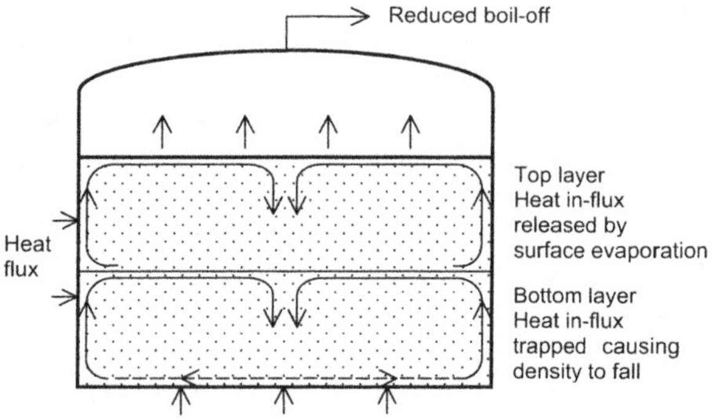


**Fig. 5.3** Stratification with low density top layer above high density bottom layer. Density difference >1.0%

Once again, the stratification is convectively stable; but now mixing across the liquid-liquid interface is controlled by double diffusion, with both temperature and concentration gradients contributing to density-gradient driven, liquid convective mixing.

This stratification in a cryogenic liquid mixture inevitably leads to unstable evaporation, which has acquired the name "rollover". The unstable evaporation takes place when the stratified layers spontaneously mix, which can lead to a rapid increase in boil-off rate and hence tank pressure.

From detailed studies of unstable evaporation from stratified liquid mixtures carried out in Tokyo, MIT, Southampton, and elsewhere [1–7], it has been observed that every experimentally simulated rollover event was spectacular and different in terms of all measurable parameters, including the time taken to spontaneously rollover, the peak evaporation rate generated, and the total vapour produced.

## 5.4 Double Diffusive Convection in Multi-component Cryogenic Liquids

In multi-component liquids, density stratification can take place in the following ways:

(1) by temperature differences between layers,
(2) by composition differences between layers,
(3) by combinations of (1) and (2).

A step-change in density across a horizontal interface between two layers will tend to disappear with time through (a) thermal diffusion across the interface tending towards the same temperature in the two layers, and (b) self-diffusion of molecular species across the interface tending towards the same composition in the two layers.

Both diffusion processes will spontaneously lead to density changes which will in turn affect convection processes and convective mixing of the two layers across the interface. Since convective mixing is much faster than diffusive mixing, it is the cross effect of diffusion on convection which is dominant.

Double diffusive convection has been widely studied from the observation of local convective motion in the sea by oceanographers, where heat and salt content are the two contributors. The standard analysis is given in, for example, references by Turner in 1965 [8] and 1975 [9], and by Huppert in 1971 [10].

Since the buoyancy or convective force on an element dm of liquid is dm g $\Delta\rho$, convection across the interface between 2 layers is governed by the sign and magnitude of $\Delta\rho$, equal to $\rho_2 - \rho_1$, the density difference across the interface between the lower layer 1 and the upper layer 2.

The density difference is given analytically from the double diffusion equation by

$$\Delta\rho = (\partial\rho/\partial T)_{x,P}\Delta T + (\partial\rho/\partial x_1)_{T,P}\Delta x_1 + (\partial\rho/\partial P)_{T,x}\Delta P \tag{5.1}$$

In the limit,

$$d\rho/dz = \underset{I}{(d\rho/dT)_{x,P}dT/dz} + \underset{II}{(d\rho/dx_1)_{T,P}dx_1/dz} + \underset{III}{(d\rho/dP)_{T,x}dP/dz} \tag{5.2}$$

For convective stability, i.e. stable stratification with no convective mixing, dρ/dz should remain negative in sign and therefore analytically less than zero. If during double diffusion across the interface, the negative value of dρ/dz becomes smaller, approaches zero, and then becomes positive, the interface becomes convectively unstable and triggers convective mixing between the two layers.

In general, term I is negative and stabilising, provided dT/dz is positive, since $(d\rho/dT)_{x,P}$ is always negative; but is positive and destabilising if dT/dz is negative.

Term II can be positive (destabilising) or negative (stabilising).

Term III is always negative and stabilising.

For cryogenic liquid mixtures, the magnitudes of terms I and II are similar, while term III is much smaller (since the compressibilities of cryogenic liquids, except helium, are small).

For most liquid mixtures, including most LNG mixtures, term II is positive and destabilising convectively against stratification. For example, preferential evaporation of less dense methane from LNG leaves a less volatile, more dense (e.g. ethane rich) surface layer which mixes spontaneously with the bulk, and there is no stratification.

On the other hand, when the more volatile (lower boiling point) component is more dense than the less volatile components, preferential evaporation of the more volatile component makes term II positive and stabilising. This will lead to spontaneous stratification, or autostratification, the consequences of which may be serious.

The most important examples of autostratification are LNG mixtures containing nitrogen, and liquid argon-oxygen mixtures (see Sect. 5.6.2).

## 5.5 Storage Behavior of Two Layers of Liquid Mixtures

### *5.5.1 The Dynamic Storage Behaviour of 2 Liquid Layers with Different Density Under Constant Isobaric Pressure*

If two multi-component liquid layers, with different densities and/or different temperatures and a common interface, are allowed to stand in a thermally isolated tank, then there is no convection; equilibration of composition and temperature takes place relatively slowly through molecular and thermal diffusion across the interface.

When heat enters the liquid space through the walls, the heat is convectively absorbed by the upward moving wall boundary layer flow which feeds into the evaporating surface. However, when there is a density reduction across the liquid-liquid interface (in the upward direction of the convection at the wall) the wall boundary flow in the lower liquid layer may have insufficient buoyancy to penetrate this density drop. Instead, the boundary layer flow turns over through 90° at the interface to become a radial inflow just below the liquid/liquid interface and then joins a central downward jet to be mixed into the lower layer. Heat entering the lower liquid via the boundary layer cannot then be released by surface evaporation, and remains trapped in the lower layer.

Looking analytically at this lack of penetration by the boundary layer flow, the quantity $(d\rho/dT)/\rho$ for LIN/LOX mixtures and LNG mixtures, is of the order of 0.5% per degree K, at their respective boiling points (see Table 5.1).

Experimental measurements on these mixtures indicate that the local temperature rise in the boundary layer is about 1 K. Thus, the buoyancy or upward force $\Delta\rho g$ on unit element driving the boundary layer flow is of the order $0.005\rho g$ and this leads to a vertical velocity in the boundary layer of the order 0.1 m/s, as is observed.

If the density drop across the interface is greater than $0.01\rho$, (or 1% of the mean liquid density) the buoyancy force driving the boundary layer upwards disappears. However, the momentum of the boundary layer will carry it beyond the interface a vertical distance of about 0.01 m, or 1 cm, before the boundary flow turns over.

This is what is observed experimentally; viz. the boundary layer flow turns through 90° away from the wall to join a radial inward flow while its vertical velocity component oscillates about the interface with an exponential decay and an initial amplitude of the order of 0.01 m.

We now have a convective mechanism whereby heat entering the lower liquid layer through the walls and floor of the storage tank cannot be released by the normal process of surface evaporation. Instead, this heat is mixed convectively throughout the lower layer, via the central downward jet.

The mean temperature of the lower layer rises, and the mean density falls with time. Eventually, the lower layer density becomes less than that of the upper layer and a convective instability is reached.

Mixing models, using double diffusive convection, are able to predict the onset of the convective instability [11] with reasonable precision, but they do not explain why the peak evaporation can be as large as 100–200 times the normal boil-off. An additional model is needed (see Sect. 5.6.2 below).

### 5.5.2 The Dynamic Storage of 2 Layers with Different Density Under Constant Isochoric Volume with Rising Pressure and Zero Boil-Off

In storage at constant pressure the stratification is unstable, as a result of surface evaporation from the top layer and heat trapped in the lower layer of an LNG mixture, leading to a decreasing density with time.

On the other hand, with storage at a constant volume under a rising pressure, with zero boil-off, the stratification into 2 layers is stable with no rollover. The top layer absorbs all the heat influx and warms up as the pressure rises, while the density rises isochorically. The bottom layer absorbs little heat and remains close to the initial temperature, while the density rises, also isochorically, as the pressure rises. The density difference between the 2 layers remains small and continues to remain small in a stable form as the pressure continues to rise. There appears to be no development with time of density differences able to promote 'rollover', unlike in isobaric storage.

## 5.6 Rollover

### 5.6.1 Basic Description of Rollover

The convective instability reached by the 2 liquid layers with different densities in the last section was originally believed to cause the stratified layers to reverse position by rolling over.

What happens in practice is that there is an unexpectedly rapid rise in evaporation rate, associated with releasing thermal overfill in the tank. If the vent valves on the tank are insufficiently sized, they will not be able to release the evaporated gas, the tank ullage pressure will rise and may lead to rupture of the tank.

Rollover is therefore a dangerous event and steps must be taken to avoid stratification building up with significant density discontinuities between layers. These steps involve the continuous mechanical mixing of the contents during all custody management operations, including filling, emptying and long-term storage [7]. In addition, the precursor of rollover, namely the development of density stratification, needs to be monitored and detected with appropriate instrumentation.

But first of all, what do we now know about the phenomenon of rollover, and how does a convective instability cause such a dramatic event?

The first recorded rollover event took place at La Spezia, and was described in detail by Sarsten in 1972 [12]. Briefly, the tank held a heel of 9500 $m^3$ of LNG with a density of 541.12 $kg/m^3$ at 114.36 K. A cargo of 33,700 $m^3$ of weathered LNG (it had been in the tanker for over a month), with a density of 544.9 $kg/m^3$ at 118.99 K, was bottom-filled into the tank with minimum mixing. The density difference was therefore 0.694%.

Eighteen hours after the filling was completed, the tank experienced a sudden rise in pressure, which marked the beginning of the rollover. This rollover event was accompanied by the evolution of a huge amount of LNG vapour lasting one and a quarter hours, with the tank vent valves open to the atmosphere. The tank took another two hours, after the vent valves had closed, to reach equilibrium with the normal boil-off rate. Some 200 tonnes of LNG were evaporated in the three and a quarter hours, with the peak boil-off estimated to be over 250 times the normal rate. Fortunately the vented gas did not catch fire, or explode.

### *5.6.2 Penetrative, Oscillating Convection Across the Interface, and Surface Evaporation Increase, During a Rollover*

In fluid dynamical terms, as the densities of two cryogenic liquid layers approach the same value with time, there comes a point at which double diffusive convection manifests itself in the following remarkable way, which has been observed about 100 times experimentally, and also recorded on video tape [13].

Since double diffusive convection is not unique to LNG or cryogenic liquids, experimental rollover simulations have been studied with LA/LIN, LOX/LIN, and liquid refrigerant mixtures, as well as LNG mixtures.

In all cases, the liquid-liquid interface is observed to become unstable whereby upwards penetrative oscillations of the interface build up in vertical amplitude. Elements of the lower layer are mixed into the upper layer, and the interface migrates slowly downwards.

Figure 5.4 shows a sequence of flow visualisation photographs, of an experimental rollover between stratified layers of R 11/113 mixtures, which illustrate the downward migration of the interface over a period of 172 min followed by it's disappearance after total mixing by penetrative convective oscillations after 182 min [5].

In the case of LOX/LIN rollovers, when the amplitude of the convective penetrative oscillations increases so that the oscillations reach the surface, the evaporation rate rises very rapidly by 10–50 fold.

The convective oscillations are believed to break up the morphology of the surface sub-layer, particularly the thermal conduction region, and cause the unimpeded and much higher molecular rate of evaporation to be realised (see Sect. 2.4.2 on molecular evaporation rates).

This high evaporation rate is then maintained so long as the convective oscillations continue to disturb the surface—up to the time when the mixing of the two layers is complete. Then the rate drops to an intermediate high value corresponding to the excess superheat temperature across the "new" surface layers of the now self-mixed (by convection) liquid of homogeneous density, but not yet homogeneous in temperature and composition. Subsequent diffusive (compositional and

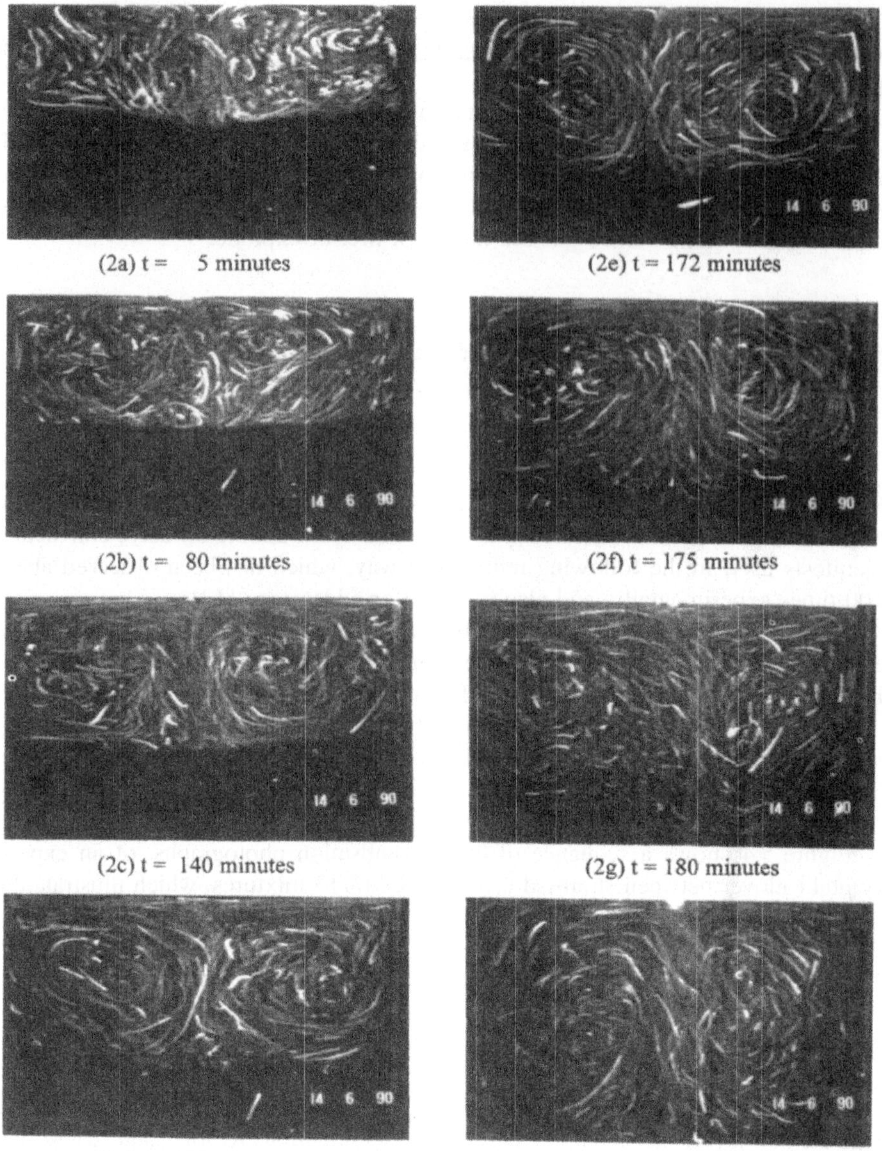

(2a) t = 5 minutes

(2b) t = 80 minutes

(2c) t = 140 minutes

(2d) t = 170 minutes

(2e) t = 172 minutes

(2f) t = 175 minutes

(2g) t = 180 minutes

(2h) t = 182 minutes

**Fig. 5.4** Sequence of flow visualisation photographs showing spontaneous mixing (rollover) of two stratified layers (with lateral and base heating) of R11/R113 mixtures differing in initial density by 1%. Exposure time was 4 s to show trajectories of seed particles thermal) mixing towards homogeneous temperature and composition is then accompanied by the intermediate high evaporation rate before the final mixture, homogeneous in temperature, composition and density, is arrived at naturally.

Optical observations of the transition through the critical point in a critical point cell (when the density of vapour and liquid phases equalise) show similar oscillatory, penetrative convection across the phase boundary.

This model, whereby penetrative convective oscillations reach the surface and break up the surface sub-layer, enables the very high peak boil-off rates to be predicted. While the estimated evaporation coefficient for LIN is $10^{-3}$, the value for LNG is not known; it could be larger by a factor of 10, thereby allowing the predicted peak boil-off rate to rise 190 fold, close to that observed in the La Spezia incident.

### 5.6.3 *Release of Thermal Overfill During Rollover*

In terms of thermal overfill, we have seen how the heat flow entering the lower liquid layer by thermal conduction from ambient through the insulation is trapped in the lower layer, causing its temperature and energy content (superheat or thermal overfill) to rise with time. The build up of thermal overfill within the lower layer is clearly triggered if, or when, the density difference between the two layers exceeds some critical value, which appears to be the order of 1.0% (0.694% for the La Spezia event).

The occurrence of two stratified layers is then the forerunner of the sequence of mechanisms leading to rollover.

When rollover takes place, it is the thermal overfill energy in the tank, not just the lower layer, which is released via the latent heat of the venting vapour. The maximum rate of generation of vapour is determined by the molecular surface evaporation limit—it is not infinitely great like an explosion.

From the 100+ instrumented, experimental rollover events made at Southampton, the following conclusions can be made:

- There was never any nucleate boiling—only surface evaporation.
- The envelope of the evaporation rate-time graph is different for every rollover.
- The total additional evaporated mass equates to the total thermal overfill energy of the two layers released during the rollover.
- Once a stratification of less dense layer on top of a more dense layer is set up, and heat energy is supplied to the lower layer, then this energy will only be released spontaneously by an inevitable rollover event.
- The peak evaporation rate may be the order of 50–250 times the normal storage rate, but it is not infinite as in an explosion.

### *5.6.4 Experimental Studies: The Two Modes or Types of Rollover*

Experimental observations on laboratory generated rollovers have been made by several groups, including those at Southampton using video recording of the flow visualisations during the rollover mixing of previously stratified layers. These studies showed that there is a whole spectrum of variations of evaporation rate with time between the extremes of two Modes of rollover. These Modes were identified by dramatically different evaporation behaviours, and by somewhat different rates of migration of the liquid-liquid interface separating the layers.

Figures 5.5 and 5.6 show summaries of two experimental runs following the setting up of two stratified layers, which demonstrated Mode 1 and Mode 2 roll-overs respectively. The difference is very significant [4].

In the first case, Mode 1, the evaporation rate increased relatively slowly with time to a peak value over a period of 60 min or more, and then subsided equally slowly back to the normal storage value. The excess evaporation tracked the rise and fall of liquid superheat in the upper layer, as the mixing proceeded spontaneously to completion and as the thermal overfill was released. The interface also moved down very slowly. All was very peaceful and controllable, without any emergency venting to the atmosphere.

**Fig. 5.5** Mode 1 rollover with LOX/LIN mixtures. Initial density difference 19 kg/m$^3$ or 2.5%. High heat flux of 67 W/m$^2$ into lower layer only

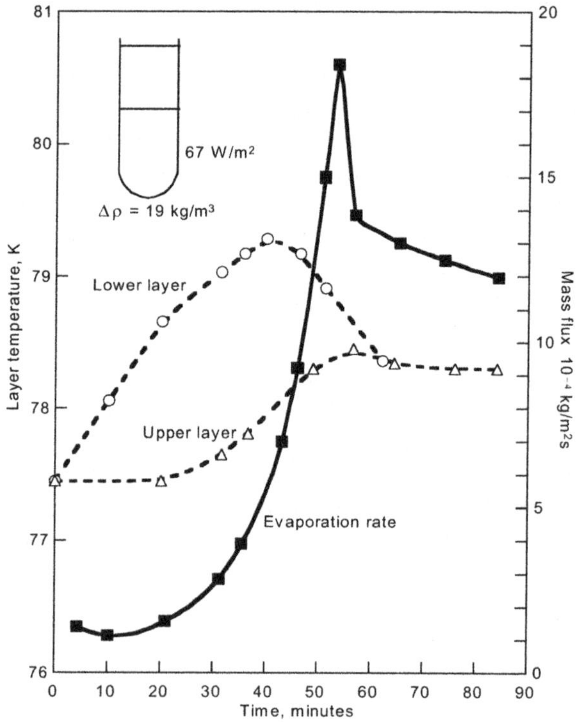

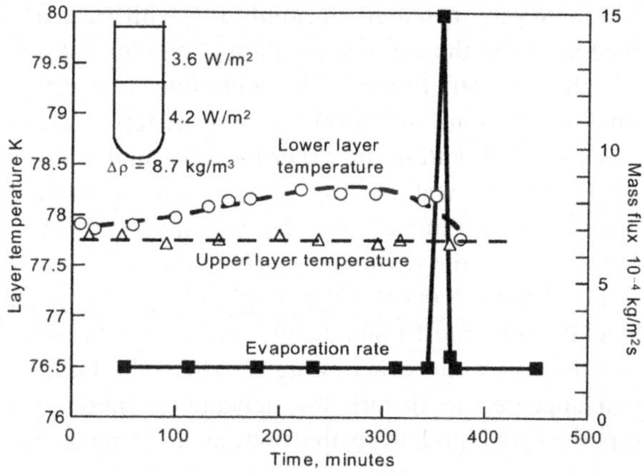


**Fig. 5.6** Mode 2 rollover with LOX/LIN mixtures. Initial density difference 8.7 kg/m$^3$ or 1%. Low heat fluxes of 3.6 and 4.2 W/m$^2$ into upper and lower layers respectively

In the second case, Mode 2, after 350 min, the evaporation rate rose very sharply within a few seconds to a high peak value, 20 times normal, before subsiding. At the same time, the interface moved downward much more rapidly than in Mode 1.

Mode 2 behaviour had features similar to that of a vapour explosion described in Chap. 3.

The studies at Southampton used binary mixtures of nitrogen and oxygen in two stratified layers and included measurements of the vertical temperature profiles, vertical composition profiles and evaporation rates, all as a function of time up to and after the rollover. However, the studies were unable to pinpoint any measurable factors which might help to predict whether the inevitable rollover would be Mode 1 or Mode 2.

### 5.6.5 *Experimental Studies: The Two Convective Mixing Mechanisms of Rollover*

There are two mixing mechanisms, reported in the literature, as the density difference diminishes with time to zero. These are:

(a) wall boundary flow penetration, and
(b) entrainment mixing via vertically oscillating convective plumes simultaneously across the whole of the liquid-liquid interface.

Both mechanisms cause upward entrainment mixing of the bottom layer into the top layer, and downward motion of the interface.

From the laboratory rollover simulations and numerical modelling at Southampton [2, 5], both wall-heating and base-heating of the bottom layer generally led to entrainment mixing by oscillating convective plumes only. The oscillating motion of the convective plumes, upwards into the top layer and downwards into the bottom layer, increased in amplitude with time as the spontaneous rollover proceeded; while the average position of the interface moved slowly down.

When the amplitude of the oscillating convective plumes built up slowly from small oscillations and never reached the top layer, then Mode 1 rollover behaviour was observed both in the experimental simulations and in the modelling (Fig. 5.7).

With slight wall-heating only, it was possible to induce wall boundary flow penetration and mixing, but this was a special case; mixing by convective plumes was generally the dominant mechanism in the experimental rollovers.

For Mode 2 events, the amplitude of the oscillating plumes increased very quickly and almost immediately penetrated through the surface sub-layer structure into the molecular evaporating region at the liquid-vapour interface. This penetration appeared to disturb the delicate morphology of the surface layers, so that superheated liquid from the bottom layer replaced the surface evaporating layer enabling the evaporation rate to rise 20–50 fold, i.e. to the molecular limit, just like a vapour explosion (Fig. 5.8).

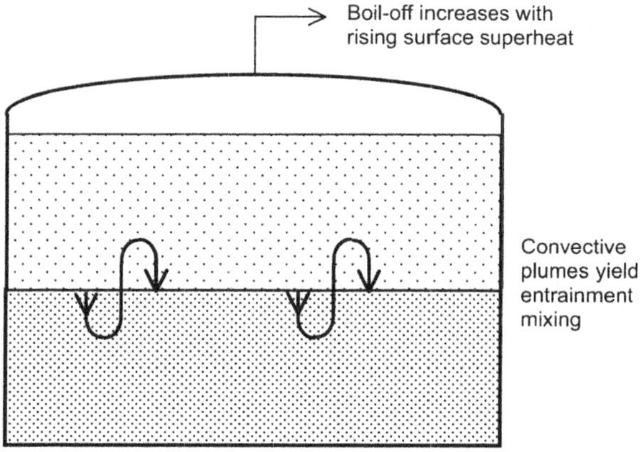


**Fig. 5.7** Mode 1 rollover

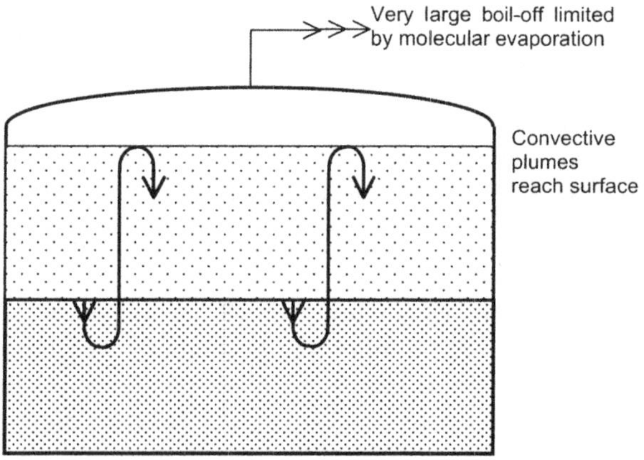


**Fig. 5.8** Mode 2 rollover

Because the entrainment mixing was so intense, the interface moved quickly downwards at the same time as the evaporation rate rose to a peak. Once the mixing was complete, the convective plumes disappeared, the disturbance of the surface layers stopped, and the evaporation rate fell relatively quickly to the intermediate value determined by the liquid superheat, as the surface sub-layer repaired itself into its normal morphology.

In some experimental rollovers. Mode 1 behaviour converted into Mode 2 when the penetrative oscillating convective plumes increased in amplitude with time and penetrated the evaporating surface sub-layer.

One feature of the numerical modelling was that the intensity of the entrainment mixing was influenced by the ratio of base to wall heat fluxes. Generally, a higher heat flux ratio led to more intense final mixing. It should be noted that in the La Spezia incident, the ratio of base to wall heat flux was high at 3.0. It may therefore be concluded that the peak boil-off during a rollover is significantly reduced if the base heat flux is no larger than the wall heat flux.

It should be additionally noted from the experiments that:

- all stratifications of two layers led to naturally occurring, spontaneous rollovers,
- the peak evaporation mass fluxes observed during Mode 2 rollovers, and some Mode 1 rollovers, could, on scaling up to storage tank dimensions, exceed the venting capacity of the safety valves on some storage tanks, and lead to structural damage.

During all the experimental studies at Southampton, at no time did the layers mix or exchange position by "rolling over". In every case, the mixing occurred via oscillating penetrative convective plumes across the whole of the liquid-liquid interface separating the two layers.

The term "rollover" is therefore not a correct description of the spontaneous mixing, but since the term is in common usage to describe this spectacular and worrying event by a single dramatic word, we shall continue to use the term also.

## 5.7 Factors Leading to Stratification and Hence Rollover

There are several ways in which a cryogenic liquid, whether a single component liquid, or a multi-component mixture, can become stratified in a storage tank.

### *5.7.1 Custody Management Creating Two Layers*

Rollover was first met in the storage of LNG, so let us consider an example assuming LNG is a two-component or binary mixture of methane and ethane. One consequence of the difference in composition between vapour and liquid is that the vapour is richer in the lower boiling point component—methane, normal boiling point 112 K, liquid density 424 kg/m$^3$—leaving the liquid richer in the higher boiling point component—ethane, normal boiling point 184 K, liquid density 544 kg/m$^3$.

It follows that, as surface evaporation continues under normal, well-insulated, storage conditions, the liquid will become progressively richer in ethane, and the density and boiling temperature, at constant tank pressure, of the liquid mixture will increase steadily with time.

Let us look more closely at an element of superheated liquid mixture at the liquid/vapour interface during surface evaporation. The spent liquid, after evaporating a methane rich vapour, has increased in density due to both evaporative latent heat of cooling and an increase in more dense ethane composition. The spent element is convectively unstable and sinks away from the surface.

Together with all the similarly spent elements, it mixes with the bulk liquid producing a uniform density and composition. There is no stratification.

Now, if a fresh supply of the same LNG binary mixture is added to the "old" or "weathered" liquid, or "heel", without being mixed with it, the fresh liquid will have a lower density and lower temperature and will tend to sit on top of the more dense heel in a very stable fashion. Thus a stratified top layer has been accidentally, or unknowingly, established.

If the density difference between layers is greater than about 1.0%, corresponding to about 5.0% difference in methane concentration, and if any difference in temperature is ignored, then the stratification will be convectively stable against mixing and will trigger the following sequence of events. This sequence has been observed in our experimental studies.

Firstly, the step in density will prevent the wall boundary flow in the lower layer from reaching the surface; the heat inflow into the lower layer will be trapped, causing the layer to heat up (Fig. 5.3). Its thermal overfill energy will increase with time and its density will decrease with time.

Secondly, and at the same time, the upper layer will be able to evaporate freely with methane-rich vapour being generated, so that the upper layer density-increases with time as the ethane concentration rises in the liquid phase.

Thirdly, the stabilising influence of the density difference between the layers will diminish with time until it approaches zero.

Fourthly, when the convective stability disappears, the layers start to mix spontaneously, or to rollover, by the penetrative mixing process. This penetration breaks up the surface sub-layer enabling the much faster molecular evaporation to take place.

Fifthly, after density equilibration and the peak evaporation has passed, the temperature difference between elements of the two layers is still large, of the order of 1–2 K. The associated thermal mixing and release of thermal overfill energy continues to generate further heating which can only be released by additional evaporation.

Sixthly, a further additional energy release arises from the heat of mixing associated with compositional equilibration, and this will add to the evaporation.

Finally, the evaporation rate decreases to its original normal value, balancing the total heat in-leak to the tank.

This sequence of events is exactly parallel to the observed sequence of events in the La Spezia incident.

Many other rollover events have been reported, or can be expected, with LNG, LPG, LIN/LOX mixtures and other industrial cryogenic liquid mixtures.

## 5.7.2 *Auto-stratification in Mixtures*

### 5.7.2.1 Mechanisms Due to Density Differences

Since stratification is the necessary precursor of a rollover event, it is important to consider the various ways in which stratification can occur, can be anticipated, and can be prevented.

We have already noted that density differences between layers can occur from temperature differences, composition differences, and combinations of the two. Also, that wall boundary layer flow cannot penetrate a density difference greater than about 1% between the layers, in which case the wall boundary layer flow is trapped in the bottom layer.

Table 5.1 shows that a 1% density difference corresponds to a temperature difference of about 1.8 K in liquid nitrogen and 2.5 K in liquid methane at their respective normal boiling points.

Likewise at constant temperature, a 1% density difference corresponds to a 2.5% change in nitrogen concentration in liquid air as a binary mixture of nitrogen and oxygen, and a 5% change in methane concentration in LNG as a binary mixture of methane and ethane.

Thus, stratification can be expected to be a common occurrence, since the necessary density changes are so small. Using the free-boiling (T-x) diagrams, the preferential evaporation of nitrogen from liquid air containing 80% nitrogen can lead to a spent liquid containing down to only 56% nitrogen with a density increase of perhaps 10%, while the preferential evaporation of methane from LNG in storage containing 90% methane leads to a spent liquid containing perhaps less than 70% methane with a density increase greater than 4%.

While it is important to realise that auto-stratification can occur in both single component liquids as well as the more usual multi-component mixtures, let us first consider several examples of auto-stratification particular to cryogenic mixtures. Then, in Sect. 5.7.3, we will consider additional examples applicable to both single component liquids and mixtures.

#### 5.7.2.2 High MW Volatile Component in the Mixture

During surface evaporation of a mixture, the more volatile, lower boiling point component evaporates preferentially, leaving a higher concentration of the less volatile, higher boiling point components in the "spent" surface layer. If the evaporating component has a density (or molecular weight, MW) greater than the higher boiling point components remaining behind, then the "spent" layer has a reducing density as surface evaporation proceeds and will spontaneously grow into an upper stratified layer, or auto-stratify.

Two important examples are LNG with a significant amount (>1.0%) of nitrogen, and liquid oxygen with a significant amount (again >1.0%) of argon. In the first case of LNG with, say, 5% nitrogen, preferential evaporation of nitrogen with MW = 28, liquid density 808 kg/m$^3$, from LNG with MW $\sim$16, liquid density $\sim$430 kg/m$^3$, would result in a liquid density reduction of 3.0%, which is more than enough for auto-stratification.

In the second case of LOX with, say, 5% argon, preferential evaporation of argon with MW = 40, liquid density 1394 kg/m$^3$, from liquid oxygen with MW = 32, liquid density 1141 kg/m$^3$, would result in a density reduction of about of 1.0%, which is probably sufficient for auto-stratification to occur.

In both examples, it should be noted that the enhanced composition of the vapour with the lower boiling point constituent, due to the surface evaporation effect discussed in Sect. 5.2, will further enhance the density reduction required for auto-stratification to take place.

#### 5.7.2.3 Non-volatile Impurities in the Surface

If carbon dioxide, water, or any other non-volatiles are present in solution, which may be at concentrations of 1 ppm or considerably less, then on evaporation, they come out of their solution phase and may subsequently remain at the surface as a monomolecular film. This film may be continuous, or it may consist of discontinuous floating rafts of impurity; but when it forms, it will provide an additional significant impedance to the surface evaporation process and literally switch off the evaporative mass flow.

A triggering mechanism has again been provided whereby the superheated liquid from the wall boundary flow cannot evaporate and a superheated layer forms immediately below the surface.

Non-volatile films may also arise from condensation of carbon dioxide, water, etc. entering the ullage vapour space during custody management operations, or from leakage of atmospheric air.

Laboratory studies with liquid nitrogen have witnessed on many occasions the switching off of evaporative mass flows for periods of many minutes, and the subsequent rapid rise in evaporation beyond the upper limit of the flowmeter. However, the introduction of carbon dioxide, either into solution in liquid nitrogen, or into the nitrogen vapour, in order to stimulate, or simulate, the formation of a film and switch off the evaporation, proved to be inconclusive. So, there are no systematic studies or results to support the idea of monomolecular films suppressing the evaporation of cryogenic liquids.

These monomolecular films do, of course, exist at ambient temperature and can be used to reduce the evaporation of water stored in reservoirs. They are also responsible for the stability of smog particles in the atmosphere.

The Schleiren photographs of evaporating cryogenic liquid surfaces sometimes show areas where there appears to be no evaporation.

#### 5.7.2.4 Marangoni Film Flow Effect

It may not be commonly known that the tank wall above the liquid level in a LNG tank is wetted by surface tension driven film flows up the tank wall. This effect, an example of the so-called Marangoni effect, was first studied at Southampton in 1974 [14]. Look into a container of LNG and see for yourself!! Then compare what you see with the tear-drops round the edge of a glass of sherry or port.

Due to surface tension differences between methane and ethane, a film of ethane-rich liquid is drawn up the wall, losing methane by evaporation, until pure ethane droplets build up at the 185 K (NBP of ethane) temperature level. The ethane droplets run back down the wall, and being more dense than LNG, they collect over a long period of time at the tank bottom as a thin, warm, dense, stratified layer. Above the 185 K level, a film of propane rich liquid is drawn up the warmer levels of the wall and propane droplets appear at the 231 K level (NBP of propane) if there is sufficient concentration in the LNG (Fig. 5.9).

However, studies of the mass dynamics of the surface tension driven separation process show that the mass flows in the films are not great enough to be commercially viable for extracting ethane or propane from LNG [15].

Another point to watch is that the wetted walls and cold exit piping, due to the Marangoni effect, make the monitoring of vapour composition particularly difficult.

### *5.7.3 Auto-stratification in Both Single Component Liquids and Mixtures*

Auto-stratification is not confined to cryogenic liquid mixtures, and can occur in a number of other ways in both single component and multi-component liquids.

#### 5.7.3.1 Self-pressurising Storage Tank

If a tank is allowed to pressurise, then the evaporation is progressively suppressed as the ullage pressure rises, and a surface layer is created with a rising temperature in equilibrium with the increasing pressure.

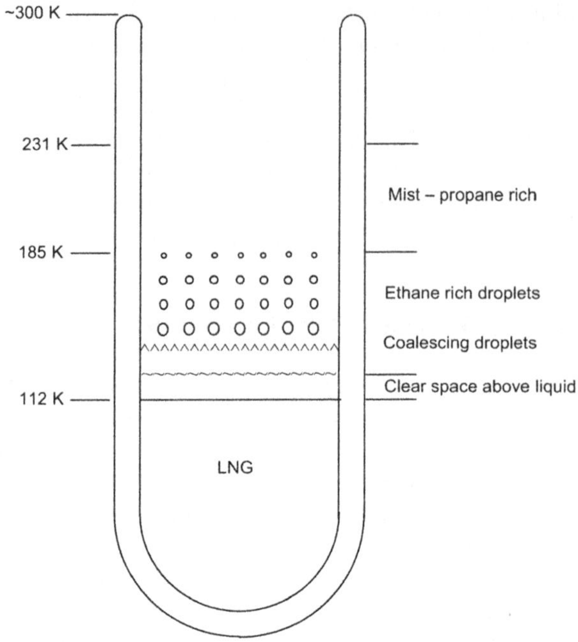


**Fig. 5.9** Marangoni effect. Schematic diagram showing appearance of transparent Dewar wall above LNG together with observed temperatures

The superheated wall boundary flow feeds into this surface without evaporating. When the density difference between bulk liquid and surface layer reaches the critical value of about 1%, the wall boundary flow can no longer feed into the surface layer, and the heat is trapped in a lower layer of bulk liquid as thermal overfill.

#### 5.7.3.2 Tall, Thin Storage Tank, Freely Venting

In a tall, thin tank, the surface area is insufficient to enable the superheated wall boundary flow, feeding into the surface sub-layer, to lose all its superheat by evaporation before entering the central downward jet.

An upper layer becomes heated by the central jet until the density difference exceeds a critical value of the order of 1%, when the wall boundary flow becomes trapped and switches into a lower layer of bulk liquid, resulting in a drop in boil-off rate.

#### 5.7.3.3 Passing Atmospheric Weather Fronts

A falling atmospheric pressure with time, from say an approaching storm front, will cause the evaporation rate to rise from a freely venting tank, and the bulk temperature to fall by up to about 1 K, until the storm front passes.

A rising atmospheric pressure with time, after say the passing of a storm front, will tend to suppress evaporation and lead to the mechanism of stratification outlined in 5.7.3.1 above. The surface layer temperature will rise to maintain equilibrium with the rising atmospheric pressure, while the bulk liquid temperature will remain constant. The associated density difference between the two layers, even for a severe storm with a pressure drop of 100 millibars, is unlikely to exceed 0.5% for most liquids, except for LHe with an estimated 2% density difference (see Table 5.1). The trapping and switching of wall boundary flow into the lower bulk liquid is therefore unlikely, except perhaps for LHe.

### 5.7.4 Custody Management Filling with Subcooled Liquid Creating Thermal Underfill

If a tank is topped up with subcooled liquid—a dangerous practice—then auto-stratification will take place immediately (see Fig. 5.10).

Subcooled liquid, i.e. liquid cooled to below its boiling point at tank pressure, will be more dense than the heel of old liquid. If mixing is incomplete, the old liquid will rise above the in-coming subcooled liquid and collect as a stratified upper layer.

The thermal overfill energy associated with this type of stratification is negative, hence "thermal underfill".

Thermal underfill is dangerous to have in a storage tank because mixing will lead to sub-atmospheric pressure in the tank and the consequential ingress of atmospheric air, or the collapse of the tank which may well not be designed to withstand a negative pressure differential.

Filling an empty tank with subcooled liquid will also lead to auto-stratification. Some of the subcooled liquid will absorb heat from the tank walls, piping, etc. to produce significant quantities of saturated liquid at the tank operating pressure,

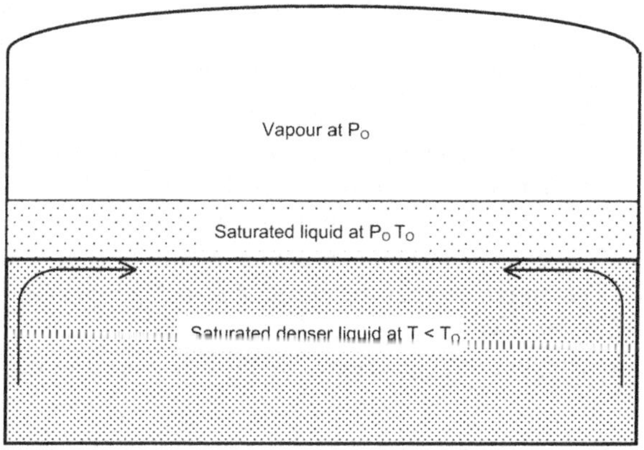


**Fig. 5.10** Autostratification by addition of subcooled liquid. For propane, when density difference between layers >1% or subcooling of lower layer >5 K. This will lead to **a** sub-atmospheric pressure in ullage space and **b** rollover with rapid increase in boil-off. Conclusion: avoid, particularly in LPG and LNG sea tankers

which partially evaporates as cooldown vapour. This saturated liquid will form a warmer, less dense stratified layer on top of the subcooled liquid.

Filling a partly filled tank with subcooled liquid may result in possibly three layers, with a top saturated layer, a middle layer of heel liquid, and a bottom layer of subcooled liquid.

Subcooled liquid must not be loaded into sea-going LNG or LPG tankers. Otherwise, all will appear to be well when the tanker sets sail. However, when the sea gets rough outside the harbour, the motion of the tanker will induce mixing and generate a negative pressure in the tanks, requiring emergency purging with nitrogen or propulsion engine exhaust gas.

## 5.8 Prevention and Avoidance of Rollover

The prevention and avoidance of stratification and inevitable rollover or vapour explosions during cryogenic storage, can be achieved by:

(1) appropriate instrumentation to detect and monitor stratification,
(2) adequate design of tank auxiliaries, fill nozzles, tank vents, vapour lines, etc.,
(3) correct custody management and early removal of thermal overfill,
(4) possible use of internal convective devices.

### *5.8.1 Detection of Stratification*

High precision instruments are required to detect density, temperature, and composition changes of the order of 1%, while location of the liquid-liquid interface requires multi-point measuring heads, or a vertically traversing head, all measuring to 0.1% precision. This level of precision is a challenge to achieve under industrial conditions.

The measurement of temperature to 0.1% can be achieved in the laboratory, but considerable effort will be required to meet this level of precision under industrial conditions; the precision of normal industrial thermometers is in the range of 1–2%.

The measurement of density with in situ instruments to the required precision of 0.1% will be difficult to achieve—the advent of cold electronics should help [16].

Measurement of composition by liquid withdrawal through a capillary tube, followed by total evaporation and gas phase analysis, is extremely time consuming and expensive for any continuous monitoring programme to be maintained.

The interface may be detectable using ultrasonic or electromagnetic (optical, radio, etc.) waves. It is therefore possible that a cryogenic sonar/radar system, using low temperature electronics, could be devised to detect stratification.

### 5.8.2 *Adequate Design of Tank Auxiliaries*

Tank systems must enable the operations to be carried out for dealing with and removing a build-up of thermal overfill through stratification, before a rollover or vapour explosion occurs. All tanks should therefore have bottom-fill capability via angled nozzle injectors, top-fill capability via spray nozzles sited well above the highest liquid level, and bottom-entry emptying (pumping) lines feeding from a sump below the bottom of the tank.

The angle of the bottom-fill nozzle injector must be such as to induce both horizontal and vertical swirl motions to the fresh liquid to achieve satisfactory mixing. The vertical swirl must be strong enough to reach the surface of liquid in the tank right up to the point when the filling operation is complete. For top filling, the spray nozzles must be widely distributed so that the fresh liquid falls evenly across the whole surface of the liquid content.

When a rollover occurs, the emergency vent valves must be large enough for the expected maximum evaporation rate not to raise the tank pressure above the design pressure. A necessary comfort for liquid management!!

### 5.8.3 *Avoidance and Early Removal of Stratification*

Stratification can be avoided by correct and adequate mixing of the tank contents, whether LNG, LPG, or liquid air, during filling operations. Mixing of fresh, lower density liquid with old weathered liquid can be achieved by bottom filling through upward angled nozzles set to induce vertical and radially inward mixing flows [7]. Instrumental monitoring is needed to test the effectiveness of filling and mixing procedures; otherwise guessing is totally inadequate and possibly dangerous.

Once stratification has occurred, the extra thermal overfill will have to be removed with some significant additional evaporation, the vapour volume generated being possibly path-dependent. This requires careful and safe tank management by mixing, either internally, or preferentially by liquid transfer between two cold tanks. If the mixing is too rapid, the additional evaporated gas flow may exceed the capacity of the vapour lines, the pressure will rise and the safety vents will open allowing large quantities of flammable gas into the environment.

If the vents cannot cope, when the flow through them approaches the velocity of sound and becomes choked with the formation of internal shock waves, the pressure will rise until structural damage occurs.

If the tank roof is lifted by the excess pressure generated, the roof may then collapse back into the liquid causing a surge of the contents, over the surviving primary containment and secondary walls, into the local environment.

This type of tank failure happened to a very large, 100 m diameter tank of LPG in the Arabian Gulf, a few years ago, with many casualties.

The practice of adding fresh liquid without mixing, so as to store two stratified layers with different density (for example, two different LPGs) in a single container or tank, is extremely dangerous, and should be stopped.

The real problem arises in deciding what to do if stratification is suspected. If nothing is done, then a rollover will undoubtedly occur.

On a tank farm with several storage tanks, one effective, and popular, procedure is to transfer liquid intermittently or continuously from tank to tank via bottom emptying and top spray filling. Passage through the spray nozzles results in some pressure drop and partial evaporation of the liquid. The liquid spray droplets have increased density and consequently sink through the top layer liquid so as to promote mixing. The additional vapour generated arises from the heat in-leak during the tank-to-tank transfer, and also from the release of thermal overfill energy arising from stratification by custody management or by auto processes.

The tank ullage pressure is determined by the rate of production of this additional vapour and can therefore be controlled, in principle, by the rate of liquid transfer. Should the ullage pressure rise too high, the liquid pumps should be stopped to try and reduce the rate of dissipation of thermal overfill energy, and allow the ullage pressure to fall back.

The safest procedure is to carry out tank-to-tank transfers continuously, when the overall rate of vapour generation should become constant. Initially, inter-tank mixing will produce a high rate of evaporation as any thermal overfill introduced during a fill with fresh, dense liquid is dissipated.

With a single tank in service, the only effective option is to circulate liquid from the bottom and in through the top fill sprays. Again, the tank ullage pressure rise is proportional to the additional evaporation produced by heat in-leak during the liquid transfer and the rate of dissipation of thermal overfill. It can again, in principle, be controlled by the rate of liquid transfer, and the pumps stopped if the pressure rises out of control. However, the circulation must be restarted, perhaps slowly at first, in order to totally remove the thermal overfill energy that has built up. There is no other option!!

If vapour is lost through the vents as a result of recirculating liquid, then this is a relatively small penalty to pay for removing a potential rollover when a much larger quantity of vapour would be lost.

Finally, one word of warning when handling a tank or vessel which is full to the brim with a cryogenic liquid: remember the significant volume compressibility of all cryogenic liquids. Under no circumstance must an excess tank pressure be controlled by suddenly switching on vapour compressors to absorb an increasing vapour flow. The vapour line pressure and hence the tank ullage pressure may be inadvertently reduced so as to cause liquid to expand and boil up into the vapour lines. The subsequent uncontrolled increase in evaporation rate and rise in pressure may lead to serious mechanical damage. Switching on vapour compressors should be carried out very gently when the vessel is brimful and then only intermittently to start with.

### 5.8.4 *Possible Use of Internal Convective Devices to Destabilise Stratification*

With a density difference of the order of 1% between stratified layers, it is possible to envisage the use of convective devices to reduce the rate of build-up of thermal overfill in the lower layer and to encourage convective mixing. Bearing in mind that once stratification occurs, the 'A' heat in-flow to the lower layer through the tank wall and floor insulation is trapped within the layer. The purpose of a convective device would then be to funnel this 'A' heat in-flow up to the top layer, thereby reducing the build-up of thermal overfill in the lower layer and delaying the onset of rollover. The question is whether the device is effective and practicable.

In tanks with a depth/diameter ratio greater than 1, the wall boundary layer suction is strong enough to pull the heated liquid in contact with the tank floor into the wall flow. Experimental attempts to increase the momentum of natural convective wall boundary layer flows in an open vessel have not been successful in the past. It is therefore unlikely that any convective device would be successful in such tanks.

In tanks with a depth/diameter ratio less than 0.5 (as in very large LNG and LPG tanks), the wall boundary layer flow may not be strong enough, and some of the heat inflow from the tank floor may be concentrated by natural convection into a number of thermals rising through the lower layer. As mentioned in Sects. 3.2 and 4.6.4, these thermals are the centres of large convection cells with horizontal spacings of the same order as their vertical dimension, namely the depth of the lower layer. Their number and position relative to the tank floor will therefore also depend on the depth of the lower layer.

If these convection thermals can be caught in large conical apertures leading into chimneys extending through the lower/upper layer interface up to the free liquid surface, then in principle the thermal catcher and chimney device could be effective in reducing thermal overfill in the lower layer [17]. The devices would have to be positioned around the tank floor and not in the centre where the central downward jet from the upper layer surface would act in opposition.

Another device includes the possible multiple concentration of the heat inflow through the floor to provide hot-spots for generating hot thermals with high buoyancy capable of passing through the lower/upper layer interface to the free surface without the aid of convective chimneys.

Devices using the introduction of gas-bubble streams are unlikely to be efficient mixers in large tanks with liquid depths of 10–50 m.

As far as we know, while all the internal convective mixing devices may work well on a small scale in the laboratory, no-one has used them in large scale storage tanks.

## 5.9 Path Dependent Mixing of Boiling Cryogenic Liquids, with Evaporation

So far in this chapter, we have not considered the volumes of vapour generated when two boiling liquids are mixed together. Let us now look more closely at the mixing process, particularly when the two liquids have widely separated boiling points.

At temperatures above ambient, the homogeneous mixing of boiling liquids is not an everyday experience. Adding immiscible liquids, like boiling oil into water, and vice versa, are probably the closest we meet, when the results are spectacularly explosive and hazardous because of homogeneous nucleate boiling of the water.

On the other hand, at temperatures below ambient, the forced convection mixing of miscible boiling cryogenic liquids, such as LPGs, LNGs, other hydrocarbon liquids, and liquid air components, is commonly carried out. Intuitively, one might expect the vapour generated to contain more of the lower boiling point component, but hardly that the volume of vapour might vary significantly with the mixing profile.

The mixing process is, of course, irreversible, and is accompanied by an irreversible increase in entropy. Mixing with evaporation is particularly difficult to model because, in addition to a possible heat of mixing, the process is significantly path dependent. A considerable volume of vapour is produced by the usual positive heat of mixing and by thermal contact between colder and hotter components (no homogeneous nucleate boiling has been observed) before the final equilibrium state of the mixture is achieved.

However, because the mixing is path-dependent, the volume of vapour produced is a variable path-dependent phenomenon. This path dependence is clearly demonstrated when liquid propane and liquid butane are mixed.

It follows that path-dependent mixing of other cryogenic liquids can be expected to produce significantly large variations in the volume of vapour generated.

### *5.9.1 Propane-Butane Mixing*

The variable vapour volume, path-dependent mixing phenomenon is, in fact, widely met in the LPG industry when mixing liquid propane. NBP −42 °C, density 581 kg/m$^3$, with liquid butane, NBP −0.5 °C, density 601 kg/m$^3$.

For example, when liquid propane is added to liquid butane, the volume of vapour generated is observed to be considerably greater, by a factor of up to 2, than when liquid butane is added to liquid propane.

Our attention was drawn to this mixing phenomenon by the problem met on board large 30,000 m$^3$, multi-tank, fully-refrigerated LPG tankers, when tank loads

(5000–10,000 $m^3$) of propane-rich liquids are required to be mixed with similar tank loads of butane-rich liquid during sea-passages. (This on-board mixing is not allowed in port).

Total re-liquefaction of the evolved vapour is required to conserve the hydrocarbon cargo, and since the on-board excess refrigeration capacity is limited, the mixing operation rate is determined by the rate of vapour evolved. At the same time, the pressures within the tanks have to be maintained within narrow margins above and below atmospheric pressure. Hence the mixing operation has to be carefully monitored so that the vapour is re-liquefied at the same rate as it is produced and so that tank working pressures are not exceeded.

When propane is added to butane, the mixing operation takes twice as long to complete because of the unexpectedly large amount of vapour generated. The transfer pumps have to be throttled, or periodically stopped to allow the mixing tank pressure to fall back to its normal range.

We carried out some experimental mixing of C3 propane and C4 butane liquids at Southampton, which demonstrated the phenomenon is not an artefact. A mixing model was then developed which showed how a correlation might be produced to predict the volumes of vapour produced [18].

### *5.9.2 Experimental Conclusions on the Forced Mixing of Propane and n-Butane*

1. Liquid propane is about 3% less dense than liquid butane and stratification is a major problem to be faced when mixing the two liquids. The (T-x) curves for propane/n-butane at 1 bar pressure are shown in Fig. 5.11 [19]. There is a wide composition difference between liquid and vapour with, for example, vapour containing 85 mol% propane in equilibrium with the liquid containing 50 mol %. The heat of mixing cannot be ignored and is calculated to be a maximum of 23 kJ/kg for a 50 mol% mixture, corresponding to the evaporation or "flash" of 5.75% of the liquid.
2. Experimental mixing results are summarised in Fig. 5.12 in terms of the percentage vapour flash, and are plotted together with the vapour flash from the estimated heat of mixing.
3. The results showed clearly that the vapour flash from adding propane C3 into normal butane n-C4 was much larger than the flash due to the heat of mixing. The vapour flash from adding n-C4 to C3 was close to the expected flash from the heat of mixing.
4. The ratio of the two vapour flash volumes was between 1.5 and 2.0. The ratio increased when the mixing time was reduced, but was largely independent of whether the propane or the butane was initially transferred to the top or the bottom.

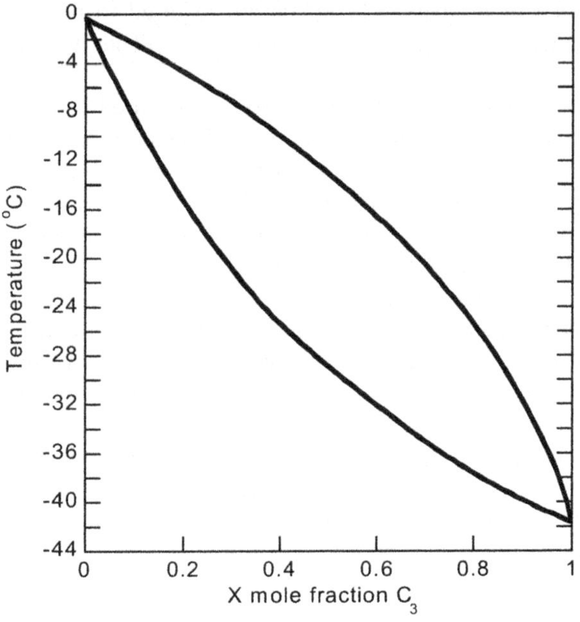


**Fig. 5.11** Temperature-composition diagram for propane/n-butane at 1 bar

5. Strangely, the vapour flash is richer in n-butane when propane is added, than when n-butane is added.
6. Experimental mixing results with commercial samples of liquid propane containing 8.5% ethane and 5.5% n-butane, and liquid butane containing 1% propane and 34% iso-butane, produced less flash, but the ratio remained about the same.
7. A path-dependent mixing model was developed which theoretically carried out the mixing by adding liquid fractions in a sequence of stages, and decoupling the vapour generation and liquid mixing processes at each stage. This model indicated how C3 added to n-C4 generates up to twice the vapour flash when compared with adding n-C4 to C3.

### *5.9.3 Some Consequences of Path-Dependent Mixing*

The experiments on propane/butane mixing have shown conclusively that vapour flash volumes are path-dependent, and may be expected to be much larger than those predicted from heats of mixing. It follows that all cryogenic liquid mixing can be expected to be path-dependent, with noticeable consequences on operations at large scales, including the following examples:

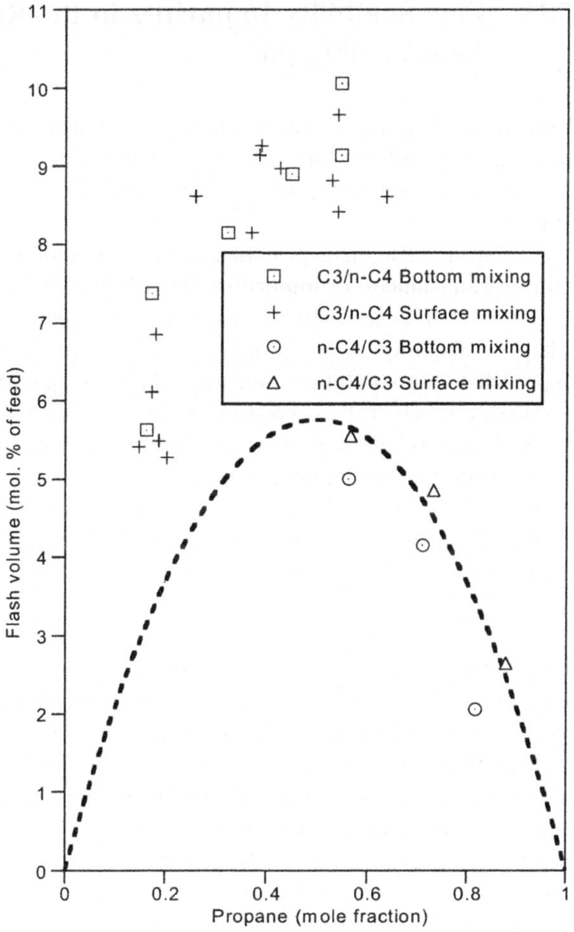


**Fig. 5.12** Experimental vapour flash results from mixing propane into n-butane, and vice versa, as flash volume versus final mixture composition. The continuous curve represents the vapour flash from the estimated heat of mixing

1. Adding and mixing fresh liquid to the existing heel, to prevent stratification. If cold is added to hotter liquid, the vapour flash volume may be greater than if hot is added to cold liquid. This applies to any pair of cryogenic liquids being mixed, whether LPG, LNG, LIN/LOX/Lair, etc.
2. When removing stratification by mechanically mixing the 2 layers, bottom to top mixing may generate more vapour flash than top to bottom mixing.
3. When a rollover occurs, the mixing is irreversible and therefore path-dependent. The penetrative mixing is from bottom to top, and the vapour flash volume can therefore be expected to be larger than predicted from the heats of mixing of the 2 layers.

## 5.10 Low Solubility Impurities in the Range 1–10 to 100 ppm

Until now, this chapter has been largely concerned with totally miscible components in a liquid mixture. However, there are many common substances which dissolve in cryogenic liquids up to relatively low limits of solubility in the 1–10 to 100 ppm range.

It is quite clear that, provided the concentration in the solution phase of these minority substances, or impurities, remains well below their solubility limit, there is no problem. Indeed, their presence will probably go unnoticed. Provided there is adequate flushing of the impurity laden cryogen through the system, then an approach to the solubility limit never happens and all is well.

However, when the cryogen is boiling as part of a separation process, such as LOX in a reboiler/condenser, and there is inadequate flushing, then a mechanism exists whereby the impurity concentration can build up and exceed the solubility limit. The impurity will then pass out of the solution phase and form a solid, (the liquid solvent is at a temperature well below the triple point of the solute), firstly as microcrystals in the liquid, and then as a solid deposit, (for acetylene in LOX, see Sect. 8.8.3).

A number of these impurity solubilities have been studied as a function of temperature, using FTIR spectroscopy and gravimetric measurements, and they all vary in the standard way as shown in Fig. 5.13 for carbon dioxide, nitrous oxide and organic compounds in liquid nitrogen, oxygen and argon [20–24].

It can be seen that the solubility decreases rapidly with decreasing temperature, falling by some two orders of magnitude between the critical temperature and normal boiling point of the solvent liquid.

In other words, a liquid under pressure and at an elevated temperature can dissolve considerably more impurity than at its normal boiling point. If a liquid solution at high pressure and elevated temperature is expanded isenthalpically through a valve to a lower pressure and temperature, the solubility limit at the lower temperature may well be exceeded and the impurity will pass out of solution.

Initially, this will be as submicroscopic crystals throughout the solvent phase, but thermo-diffusiophoresis in local temperature gradients downstream of the valve will aid mechanical deposition on to the walls of the pipework or vessel, where the impurity will collect. The submicron crystals are unable to grow because the temperature is too low, and they remain as a finely powdered solid with a very large surface area/volume ratio.

Some of these cold, finely powdered solid impurities are benign, e.g. water and $CO_2$, but some finely powdered solid impurity/cryogen combinations may be pyrophoric or spontaneously combustible in the cryogenic liquid. The latter represent a safety hazard and are discussed in Chap. 8 on Safety.

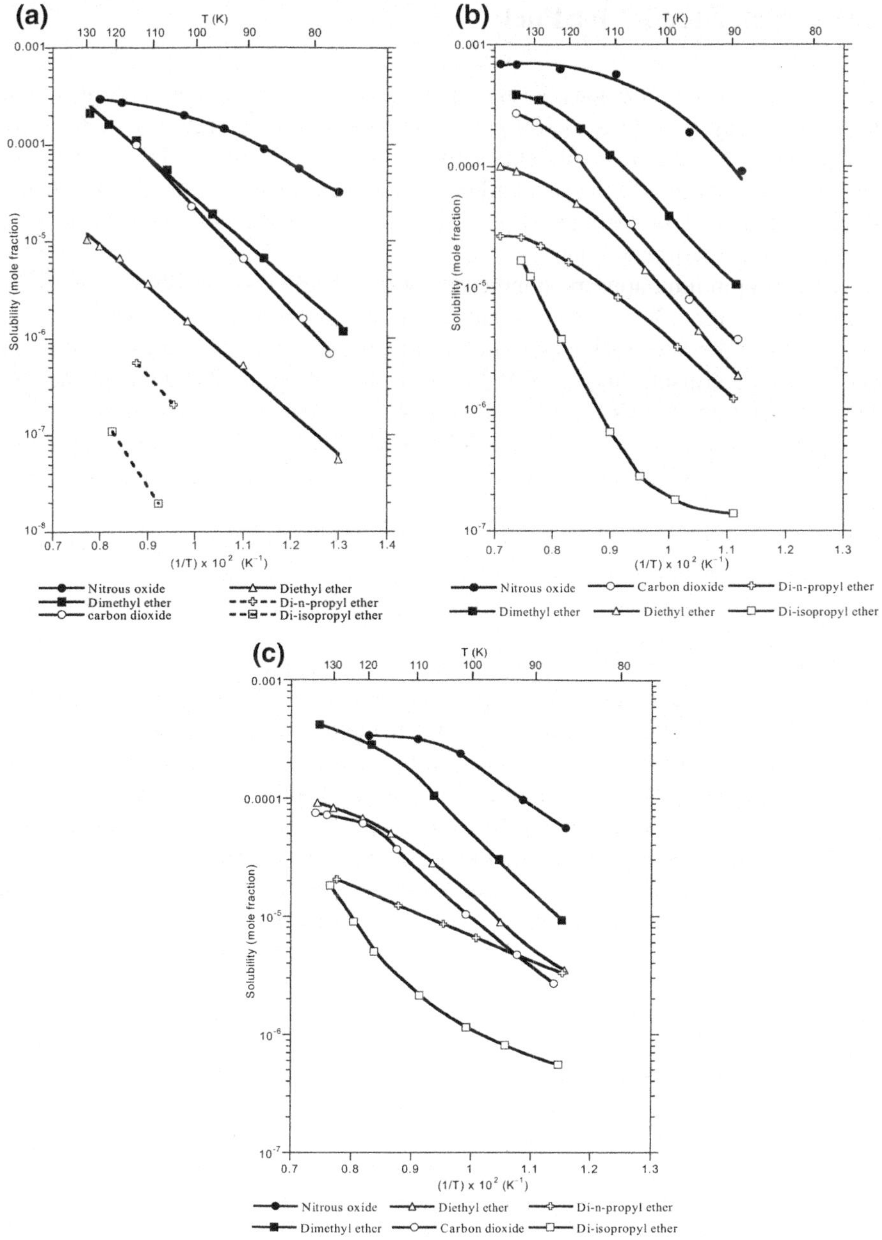


**Fig. 5.13** **a** Solubility of solutes in liquid nitrogen as a function of temperature. **b** Solubility of solutes in liquid oxygen as a function of temperature. **c** Solubility of solutes in liquid argon as a function of temperature

## 5.11 Water/Ice in Jet Fuel

As a mixture of hydrocarbons, jet fuel may absorb water from the atmosphere and other sources up to a limit of 100 ppm. The variation in solubility of water in jet fuel is similar to other low solubility impurities, and decreases rapidly with falling temperature according to the standard cryogenic relation, log s = A/T + B where s is the solubility at temperature T and A and B are constants which have to be measured by experimental data.

The 100 ppm limit appears to indicate water will come out of solution below the temperature range −35 to −45 °C, and freeze into small particles of ice. Now, the wing tanks of the latest passenger aircraft, like the Boeing 777, are exposed to the cold air at the cruising height of 40,000 ft at the bottom of the stratosphere. The static temperatures at this level are in the range −65 to −75 °C, and after allowing for temperature rise from skin friction, the tank walls/wing surface may be as cold as −50 to −65 °C depending on the unpredictable temperatures in the lower stratosphere.

If the solubility limit is exceeded at these temperatures, then the dissolved water will freeze out as ice on the inside of the tank walls (see Fig. 5.14).

Figure 5.14 Predicted rapid fall in solubility of water "s" in jet fuel, with decreasing temperature.

When the aircraft descends into warmer air towards the end of the flight, the ice starts to melt and becomes detached from the tank wall. The ice particles may then start to block the fuel lines and heat exchangers thereby restricting the flow to the jet engines. This is believed to have happened to a Boeing 777, which crash landed at London Heathrow when both engines failed on its landing approach.

The Air Accident Investigation Branch accepted the cryogenic explanation and therefore recommended that all high flying passenger jets should start descending at least 30 min before landing: they should also descend to warmer air in the top of the troposphere at 30,000–35,000 ft if the static air temperature is observed to fall

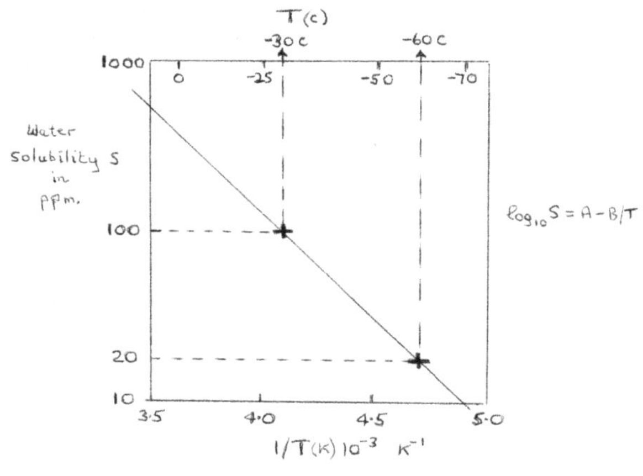


**Fig. 5.14** Solubility of water in jet fuel. Predicted rapid fall in solubility "s" of water in jet fuel, with decreasing temperature

to −70 °C. The result of these recommendations, which have been accepted by all aircrews, is that no further engine failures due to ice in jet fuel have been reported, to the time of writing (see Ref. [25]).

## 5.12 Summary on Mixtures

1. (T-x) data for surface evaporation of mixtures may not be the same as the (T-x) free-boiling data used for distillation. The deviation in vapour composition from the equilibrium data is proportional to the surface evaporation mass flux.
2. While stratification in a single component liquid may lead to QHN boiling, the consequences of density stratification are more common in multi-component liquid mixtures. The density varies with both temperature and composition, and stratified layers experience double diffusive convection instabilities and spontaneous mixing or rollover.
3. The inevitable consequence of stratification in a mixture is a rollover incident when the 2 layers mix spontaneously by vertical penetrative oscillating convection across the whole of the liquid/liquid interface.
4. The peak boil-off during rollover is determined by:

   (a) the increased superheat of the surface layer in Mode 1 rollover, or
   (b) the penetrative convection reaching the surface sub-layer in Mode 2 rollover, and breaking down its morphology so that the full unimpeded mass flux of molecular evaporation is reached at 20–250 times the normal boil-off, depending on the magnitude of the evaporation coefficient.

5. There are several ways in which autostratification can occur.
6. Monitoring of tanks for the occurrence of stratification is necessary, otherwise continuous pumping of the liquid should be applied to promote mixing as a preventative measure.
7. Impurities with limited solubilities, which all decrease rapidly with decreasing pressure and temperature, present problems through coming out of solution during long term storage or transfer of impure liquids.
8. This summary applies to all cryogenic liquid mixtures, whether LPG, LNG, jet fuels or LIN/LOX/Lair mixtures.

## References

1. Agbabi, T., Beduz, C., Scurlock, R.G., Shi J.Q.: Evaporation stability of cryogenic liquids under storage. In: Proceedings of LTEC 90, 1.5 (1990)
2. Shi, J.Q., Beduz, C., Scurlock, R.G.: Numerical modelling and flow visualisation of mixing of stratified layers and rollover in LNG. Cryogenics **33**, 1116 (1993)
3. Scurlock, R.G.: Stability of cryogenic liquids under storage. In: Proceedings of Kryogenika 90, Kosice (1990)

4. Agbabi, T.: Rollover and interfacial studies of LNG mixtures. Ph.D. thesis, Southampton University (1987)
5. Shi, J.Q.: Numerical modelling and experimental study of rollover. Ph.D. thesis, Southampton University (1990)
6. Sugawara, Y., Kubota, A., Muraki, S.: Rollover test in LNG storage tank and simulation model. Adv. Cryog. Eng. **29**, 805 (1983)
7. Smith, K.A., Lewis, J.P., Randall, G.A., Meldon, J.H.: Mixing and rollover in LNG storage tanks. Adv. Cryog. Eng. **20**, 124 (1974)
8. Turner, J.S.: *Buoyancy effects in fluids*. Cambridge University Press, Cambridge (1979)
9. Turner, J.S.: The complex turbulent transport of salt and heat across a sharp density interface. Int. J. Heat Transfer **8**, 759 (1965)
10. Huppert, H.E.: On the stability of a series of double-diffusive layers. Deep Sea Res. Oceanogr. Abstr. **1005**, (1971)
11. Germales, A.E.: A model of LNG roll-over. Adv. Cryog. Eng. **21**, 330 (1975)
12. Sarsten, J.A.: LNG stratification and roll-over. Pipelines Gas J. **199**, 37 (1972)
13. Beduz, C., Scurlock, R.G.: Spontaneous convective mixing or "rollover" between two stratified layers of LIN/LOX mixture. Video tape demonstrated at Heat Transfer Conference, San Francisco (1986)
14. Booth, D.A., Bulsara, A., Joyce, F.G., Morton, I.P., Scurlock, R.G.: Wall film flow effects with LNG. Cryogenics **14**, 562 (1974)
15. San Roman, O.: The dynamics of methane/ethane separation by differential surface tension driven flows; or the Marangoni effect. Ph.D. thesis, Southampton University (1978)
16. Scurlock, R.G.: On-line instrumentation of cryogenic systems and plant to an accuracy of 0.01% using cold electronics. In: Proceedings of ISA, Houston, **32**, 139 (1993)
17. Voyteshonok, V.: Safe storage vessels for low-boiling liquids. Cold Facts **21**, 33 (2005)
18. Tchikou, A.: Mixing of propane and butane. Ph.D. thesis, Southampton University (1985)
19. Williams, A.F., Lorn, W.L.: Liquefied Petroleum Gases. Ellis Horwood (1982)
20. Rest, A.J., Scurlock, R.G., Wu, M.F.: Criteria for assessing the solubility of solutes in cryogenic liquids. Cryogenics **25**, 591 (1985)
21. Rest, A.J., Scurlock, R.G., Wu, M.F.: The solubilities of $N_2O$, $CO_2$, aliphatic ethers, alcohol, and water in cryogenic liquids. Chem. Eng. J. **43**, 25 (1990)
22. Rebiai, R.: Solubility of non-volatile impurities in cryogenic liquids. Ph.D. thesis, Southampton University (1985)
23. Wu, M.F.: Solubilities in cryogenic liquids. Ph.D. thesis, Southampton University (1986)
24. Yun, S.: Phase equilibria for $CO_2$ in LNG components. Ph.D. thesis, Southampton University (1988)
25. Scurlock, R.G.: Cryogenic problems of flying long-haul at 40,000 ft. (12,200 m) in stratosphere at −57 °C. In: Proceedings of IIR Cryogenic Conference, Prague (2010)

# Chapter 6
# The Handling and Transfer of Cryogenic Liquids

## 6.1 General Remarks on Subcooled Liquids and 2-Phase Flow

We are all used to handling and transferring water, whether we are using a hosepipe to wash the car, filling a kettle, or turning taps to have a bath or shower. In all cases, whether the water comes from the cold tap at ambient temperature or from the hot tap at, say, 60 °C, the water is subcooled way below its normal boiling point. It is subject to a pressure, whether it is atmospheric pressure or a greater hydrostatic pressure, which is well above its saturation vapour pressure. The water transfers are never a problem because there is no 2-phase flow—unless dissolved air comes out of solution, when there may be a hint of problems like water-hammer, irregular and reduced flow, and so on.

Handling and transferring petrol is not quite so easy because the normal degree of subcooling is not so large as with water. Storing petrol at sub-ambient temperature in underground tanks at filling stations ensures there are no 2-phase problems when filling a road vehicle's fuel tank. However, a vapour lock in a hot engine cutting out the fuel supply is a good example of the sub-cooling being lost in the fuel supply line, thereby creating a 2-phase problem.

Handling cryogenic liquids is just as easy as handling water or petrol if the liquids are adequately subcooled from the start, right through to the end of any transfer process. However when 2-phase flow occurs, the problem is a major one and the transfer may slow down, or stop altogether.



T. D. Bostock and R. G. Scurlock, *Low-Loss Storage and Handling of Cryogenic Liquids*, International Cryogenics Monograph Series,
https://doi.org/10.1007/978-3-030-10641-6_6

## 6.2 What is 2-Phase Flow?

When any liquid evaporates, whether cryogenic or otherwise, a very large volume of vapour is generated at the same pressure. In the case of liquid helium or nitrogen, the volume of vapour at ambient temperature and pressure is some 700 times the volume of the liquid before it evaporates. This means that, say, if only 1% of the mass of the liquid in a transfer line evaporates, the volume occupied by the vapour is 7 times the total volume of unevaporated liquid. Although the vapour mass is 99 times less than the liquid mass, the vapour volume occupies seven eighths of the volume of the line, and the mass flow-almost stops.

In general, a mixture of vapour and liquid, having a much lower density, must have a much greater velocity in order to maintain a required mass flow rate. The limiting velocity is, of course, the local velocity of sound in the fluid. While the velocity of sound in cryogenic vapours is quite high ($\sim$300 m/s for nitrogen and oxygen, 1000 m/s for helium, 1300 m/s for hydrogen at 300 K, and proportional to $T^{0.5}$), it is much lower in any 2-phase mixture because of the high adiabatic compressibilities of all mixtures.

Useful empirical correlations have been developed by Martinelli and Lockhart [1] for calculating frictional pressure drops at ambient temperature, which have been extended with reasonable accuracy to low temperature flows [2, 3]. The net result is that the occurrence of 2-phase flow with a fixed available overpressure for liquid transfer, will lead to a significant fall in mass flow and a mass transfer rate close to zero.

Likewise, if 2-phase flow develops in a rotary liquid transfer pump, the mass flow output will be reduced or the pump may fail to prime altogether, leading to overheating and mechanical failure. Transient or continuous oscillations may also lead to mechanical damage.

The obvious step is to avoid 2-phase flow but, first, there is a need to understand how 2-phase flow happens.

## 6.3 Occurrence of 2-Phase Flow

Let us consider the P–T diagram in Fig. 6.1, where XY is the saturation vapour pressure–temperature line, or alternatively the pressure versus boiling point curve. In addition to showing the saturation vapour pressure curve, the diagram can also be regarded as a thermodynamic state diagram, with the curve separating liquid and vapour phase thermodynamic states.

- All points above the curve represent liquid phase states,
- all points below represent vapour phase states,
- and all points on the curve represent liquid and vapour in contact with one another, namely the 2-phase state.

**Fig. 6.1** P–T diagram showing saturation vapour pressure–temperature curve separating liquid and vapour thermodynamic states

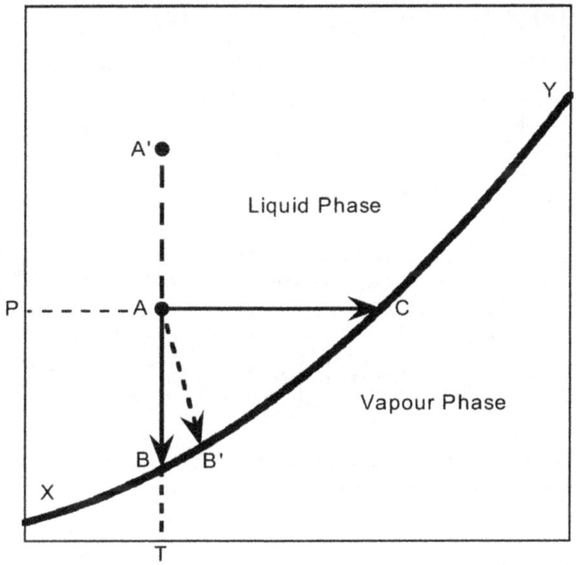


Consider liquid with thermodynamic state 'A' at pressure P and temperature T. The liquid is undercooled, or subcooled with respect to its boiling point at pressure P.

During a transfer operation, the liquid state can cross the saturation vapour pressure curve XY resulting in the creation of two phases by two separable paths, or by combinations of the two paths, i.e.

(1) by a reduction in pressure along AB (or strictly along AB′),
(2) by absorption of heat, producing a rise in temperature at constant pressure, along AC,
(3) by various combinations of pressure reduction and heat absorption.

To prevent the occurrence of 2-phase flow during a liquid transfer, the change in thermodynamic states represented by both paths AB and AC must not end on, or cross, the saturation vapour pressure curve XY.

This can be simply achieved with adequate subcooling of the liquid, by pressurising the liquid along AA′ before the transfer commences.

## 6.4 Pumped Liquid Transfer Avoiding 2-Phase Flow

Consider Fig. 6.2a where we have a storage tank containing liquid, density ρ, depth L, stored at pressure $P_0$ and saturation temperature $T_0$ with a pump mounted a vertical distance H below the bottom of the tank. Liquid entering the pump is subcooled by the hydrostatic pressure head ρg (H + L), but loses some pressure

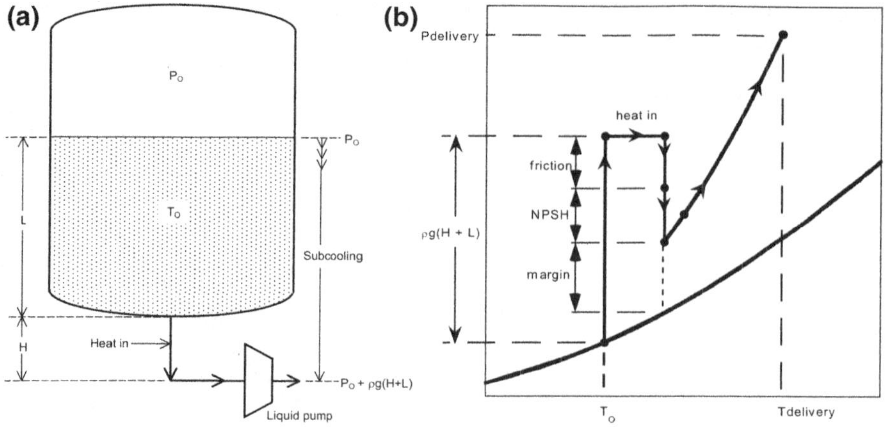


**Fig. 6.2** **a** Typical pumping facility. **b** P–T diagram showing change in thermodynamic state of liquid during a pumping operation

through frictional pressure drop and gains heat through the wall of the connecting line. The changes in thermodynamic state are shown in Fig. 6.2b.

Like any liquid pump, whether rotating or reciprocating, the Net Positive Suction Pressure Head (NPSH) is a primary requirement. Clearly, the reduction in liquid pressure as it is sucked into and accelerates through the inlet passages of the pump, namely the NPSH, must not cause the liquid state to cross the saturation line into the 2-phase state. If it does, then cavitation or partial evaporation will occur causing the pump to stall. The liquid pressure before entry to the pump must therefore always be greater than $P_0$ by the magnitude of the NPSH for the pump.

As the pumping process continues, the liquid level falls reducing the hydrostatic pressure head and associated liquid subcooling, at the entry to the pump. Eventually, the total pressure head falls to the NPSH or below; the pump stalls and will no longer remove further liquid from the tank.

This is not the whole story, because, at the same time, the ullage pressure in the vapour space is falling. As the liquid level drops, the vapour in the ullage space expands into the volume previously occupied by the liquid removed. Generally, there is not enough liquid evaporation to maintain the ullage pressure, and the pressure falls, causing the liquid subcooling at the pump entry to further reduce. This latter effect can be alleviated by back-filling vapour into the tank.

When the tank is too large to be vacuum insulated and the insulation is provided by a gas purged foam or powder at 1 bar, it is desperately important that a negative pressure is not generated inside the tank by the liquid pumping process. Back-filling with vapour flashed from liquid in the pump delivery line should be employed. Otherwise, the negative pressure could lead to the tank-wall collapsing inwards.

## 6.5 Liquid Transfer Techniques Avoiding 2-Phase Flow

Starting with a stored liquid at pressure $P_0$ and saturation temperature $T_0$ there are a number of methods for creating the necessary positive pressure differential and liquid subcooling for transfer as follows:

(1) Pressurising by external use of a non-condensing gas. For example, helium is used for pressurising LOX and liquid hydrogen space rocket tanks.
(2) Pressurising by external use of the same gas. The introduction of warm gas at a pressure greater than $P_0$ into the ullage space will rapidly raise the pressure. Some condensation at the liquid surface takes place so as to form a thin layer of saturated liquid in thermodynamic equilibrium with vapour at the new pressure in the ullage space. Stratification prevents this surface layer mixing with the rest, and little further condensation of the pressurising gas will take place.

While this technique is applicable to larger scales, another technique is available for laboratory scale systems using liquid helium or hydrogen. Adequate pressurising is achieved by the use of a rubber bladder connected to the vent line of the storage vessel from which the liquid is to be transferred. Squeezing the bladder forces warm gas down the neck of the vessel causing a small quantity of liquid to evaporate and thereby generating the excess pressure necessary for liquid transfer.

(3) Pressurising by self-heating, using the normal heat inflow through the insulation. As discussed previously, this heat flow will generate a warm stratified upper layer, with its surface in equilibrium with the higher pressure in the ullage space. Transfer of liquid will be accompanied by a continuous fall in pressure with a diminishing transfer rate.

This is because the transfer introduces motion into the whole of the liquid contents and causes subsequent mixing of the warm upper stratified layer into the colder lower layers. The surface temperature falls, there is some condensation of vapour and the ullage pressure also falls. At the same time, expansion of vapour into the volume previously occupied by the exiting liquid also contributes to the fall in pressure.

Thus, self-heating is adequate for transferring relatively small quantities of liquid from an efficiently insulated tank, say 5–10% in a period of 24 h.

This is more than sufficient for filling 25 l laboratory transfer Dewars from a 10,000 l vacuum insulated storage tank of LIN.

(4) Pressurising by use of a pressure raising coil or vaporiser connected between bottom and top of the vacuum-insulated tank. The flow is driven by the difference in hydrostatic pressure heads between liquid in the tank and vapour in the line above the vaporiser, and may be simply controlled via a valve in the vapour line to the top of the tank. Automatic control via a pressure regulator enables the tank ullage pressure to be maintained independently of the rate of liquid removal.

(5) The use of single stage and multi-stage mechanical pumps, generally rotary, for liquid transfer purposes. These can be:

   (a) submerged in the liquid, are permanently cold and therefore ready for immediate transfer, e.g. for LIN and LNG, or
   (b) externally mounted, and require cooling and priming before transfer can take place, e.g. for LOX.

In all cases, the liquid transfer will stop, or perhaps not even start, if the liquid state crosses the saturation curve creating 2-phase flow at any point in the transfer process.

The consequences of a transfer failure can vary from just an inconvenience, resulting in wasted time and cryogenic liquid, to a major fire and explosion in a mechanical pump which overheated when it failed to prime. It is therefore important for operators of transfer equipment to understand the importance of avoiding 2-phase flow by working with an adequate margin of subcooling.

## 6.6 Liquid Transfer with Transient 2-Phase Flow

During the cooldown of liquid transfer lines and storage tanks, vessels or cryostats, the occurrence of 2-phase flow, albeit transient, is unavoidable.

Now, in steady state 2-phase flow, there are a variety of ways in which vapour and liquid can co-exist in a pipeline, such as slug, mist and annular flows, all of which are well documented in texts on boiling heat transfer and cryoengineering (see Sect. 6.2).

During a cooldown process, there are three flow sections:

(a) liquid in the cold portion of the transfer line,
(b) 2-phase flow and cold vapour flow where cooldown is taking place,
(c) warm vapour exiting through the vapour lines.

The pressure drop in (a) the length of line filled with cold liquid is relatively small and can generally be ignored.

The pressure drop in (b) the cooldown area is higher but need not be excessive if the flow is not being forced. The heat transfer process which is achieving the cooldown is a mixture of natural and forced convection heat transfer between cryogen and solid surfaces of the cryogenic system and is difficult to model accurately. The main finding is that the heat transfer is only weakly dependent on cryogen velocity and hence pressure drop. Forcing the flow with a high pressure input will not noticeably increase the rate of cooldown.

The pressure drop in (c), the vapour venting section, is by far the largest quantity. This is because the volume flow of warm cryogen vapour is much greater than the volume flow in the 2-phase cooldown section, and is several hundred times greater than the liquid input volume flow.

The best way of understanding cooldown is to consider some examples, namely a long transfer line, a cryostat, a tank and a large mass such as a superconducting magnet.

## 6.7 Cooldown of a Long Pipeline with L/D Greater Than 2000

With a long pipeline, length L, inner diameter D, the question is how to cool the pipe down to a liquid temperature and commence transfer, the operation to be as fast, and as thermodynamically efficient, as possible.

Precooling a transfer line with a second liquid to an intermediate temperature, e.g. LIN precooling for a helium line, is not an effective option even if the liquid is conveniently available. It is time consuming and it is also difficult to remove all traces of the second liquid. For example, all traces of LIN must be removed from the line before LHe is admitted to complete the cooldown. Otherwise, any solid nitrogen produced will require a great deal of LHe to cool it down to 4 K (because solid nitrogen has a high heat capacity and poor thermal conductivity), even if it does not block the line.

The simplest and most effective method is to use the liquid cryogen to be transferred. In addition to the latent heat of evaporation, there is the all-important sensible heat of the vapour from boiling point up to ambient temperature available for cooling the pipeline. As mentioned several times already, this sensible heat is the same magnitude as the latent heat of evaporation for nitrogen, oxygen and natural gas, is 8 times the latent heat of hydrogen and 75 times the latent heat of helium.

Using subcooled liquid from one end, at an entry pressure above saturation pressure, a "cooldown wave" progresses along the pipeline as indicated in Fig. 6.3.

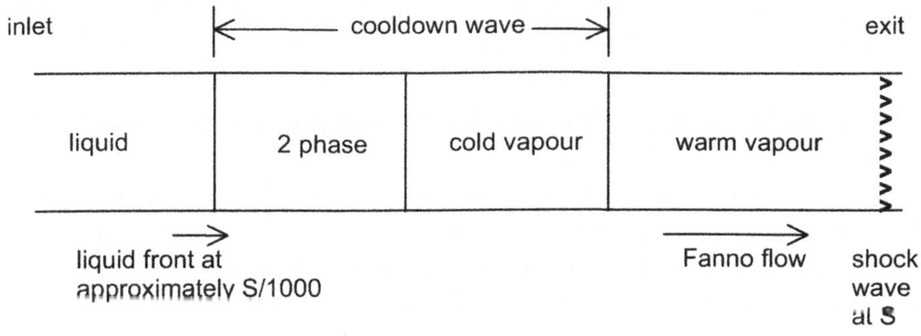


**Fig. 6.3** Cooldown of pipeline showing liquid front, cooldown wave, Fanno flow of warm vapour, and shock wave at the exit (S = local speed of sound)

Ahead of the liquid front, 2-phase flow occurs as the inner wall of the pipeline, inner diameter D, is cooled down to the liquid temperature; then there is single phase cold vapour cooling the pipeline to a point about 1000 D further down, where the temperature difference between wall and vapour has decreased to zero. The region from this "warm" point back to the liquid front is the so-called "cooldown wave".

Beyond this warm point, the wall and vapour are both at approximately ambient temperature.

The controlling feature of the entire cooldown process is the behaviour of the ambient temperature vapour, as it accelerates down the uncooled portion of pipeline until it reaches the local speed of sound S near the exit. A Shockwave develops and all the remaining pressure drop along the pipeline is consumed by increasing the intensity of the shock wave and not by increasing the mass flow. This flow, or Fanno flow, is well-known to blow-down wind-tunnel operators, in which the maximum mass flow is determined only by the local speed of sound S at ambient temperature, the exit pressure and the cross-sectional area of the exit (see for example, Shapiro [4]).

The velocity of the liquid front behind the cooldown wave is then determined entirely by the Fanno flow of the ambient temperature vapour flow ahead of the cooldown wave.

In addition to cooling the inner wall of the pipeline, the evaporating liquid has to absorb an increasing heat flow through the surrounding insulation as the liquid front advances. The liquid front velocity v is therefore considerably less than the figure of $S \times \rho(\text{vapour})/\rho(\text{liquid})$, i.e. less than S/1000, where ρ(vapour) and ρ(liquid) are the vapour density at ambient temperature, and liquid density respectively.

For nitrogen and helium at 300 K, S is about 300 and 1000 m/s respectively; hence v is less than the order of 0.3 and 1.0 m/s respectively, reducing significantly as the liquid front progresses down the pipeline.

It should therefore not be surprising that it can take a long time to pre-cool a long pipeline before liquid can be transferred. Increasing the inlet pressure will not increase the mass flow rate or speed up the rate of cooldown; the exit shock will just get stronger.

It follows from these points that all transfer lines should be as short as possible.

## 6.8 Cooldown of a Cryostat with Minimum Loss of Liquid

To cool a cryostat down to liquid temperature efficiently, full use should be made of the "cold" in the cryogen, including both the latent heat of evaporation and the sensible heat of the vapour between boiling point and ambient temperature to cool the system from the bottom upwards. This can only be achieved if:

(1) the liquid is fed into the very bottom of the vessel, because the evaporated vapour will only flow upwards by its buoyancy inside the vessel,

(2) the vapour passes out of a vent exhausting from the top of the vessel, and
(3) the liquid is fed in slowly under low pressure difference.

The heat transfer process between cold vapour and vessel wall is by natural convection, in which the heat transfer coefficient is (a) small and (b) almost independent of vapour velocity. The rate of cooldown cannot therefore be accelerated by having a high velocity vapour stream.

In other words, the cooldown is a leisurely process which cannot be hurried—if you try to hurry by increasing the flow rate, liquid will only be wasted. A sound indication that you are trying to cool down too quickly is the build-up of frost on the vent pipe.

I have demonstrated liquid helium cool downs to my students on many occasions with **no** frost on the vapour vent pipe, by going easy on the cooldown flow rate and using a small pressure differential—and cooling down just as quickly as with transfers under high pressure differential with frost over all the pipework and using about 5 times as much liquid helium.

The secret to achieving an efficient cooldown, with minimum use of cryogenic liquid, is to let nature and natural convection take their course and then there is no stress!!

## 6.9 Cooldown of a Tank

A large tank may be provided with an overhead liquid spray near the top as well as a bottom liquid inlet. The overhead spray should not be used for cooling down the empty, warm tank since the liquid spray droplets will be carried out via the vent and their cooling capacity will be lost. The spray is intended for mixing a fresh supply with an older heel of liquid already in the tank and preventing stratification between the two liquids.

The most efficient cooldown is ensured by admitting liquid via the bottom fill inlet only.

## 6.10 Cooldown of a Large Mass Such as a Superconducting Magnet

As with a pipeline, precooling with a second liquid such as LIN, prior to admitting liquid helium, is a doubtful option, because of the problems of removing all traces of the second liquid. It is easier, and far less time-consuming, to use a single liquid cryogen to cooldown from ambient by:

(a) ensuring that all the available enthalpy of the liquid cryogen is used, both latent heat and sensible heat throughout the cooldown by using a bottom fill procedure and,
(b) being aware that the relatively low rate of heat transfer between cold vapour and hotter solid is a fact to be accepted.

To start the cooldown, a relatively high pressure differential is needed because there is a large back pressure from the escape of warm vapour up through and around the superconducting magnet. However, as the cooldown proceeds, the back pressure will decrease and the liquid transfer rate will tend to rise. The rate of heat transfer will not rise and it will be necessary to reduce the pressure differential to ensure full use of the available enthalpy of the liquid and cold vapour. Otherwise, there will be a build-up of frost on the vent line indicating that the available cold is being wasted.

There may well be other reasons for slowing down the cooldown, for example, the need to maintain small temperature gradients within the magnet to keep down mechanical stress levels arising from differential contractions.

## 6.11 Insulation of Transfer Lines

The degree of insulation depends on the duty cycle and cryogen to be transferred. For liquids above 81 K, the condensation temperature of air, it could be that more liquid will be lost in cooling down the insulation than with no insulation at all. This applies to the flexible connections used for filling storage tanks from road and rail tankers, particularly with LOX, LA, and LNG.

This also applies, in particular, for LIN delivered at pressures above 2 bar and temperatures above 81 K, but will require the LIN to be flashed down to a lower pressure at the discharge end (but see next Sect. 6.12).

## 6.12 Flashing Losses Due to Transfer at Unnecessarily High Pressures

Any pressure reduction at constant enthalpy during transfer of saturated liquid will result in significant loss of liquid by flashing. For example, if LIN at 100 K, 8 bar (120 psi) is reduced to 77 K, 1 bar (15 psi) by expansion through a control valve, then 25% of the liquid is evaporated and lost as a flashing loss during the transfer.

Customers object to losing 25% of the liquid they are buying because the tanker is delivering at an unnecessarily high pressure—2 bar should be an adequate tanker pressure, causing the loss from flashing to 1 bar to be reduced to less than 5%.

## 6.13 Zero Delivery

As the liquid front at the cold end of the cooldown wave advances along the delivery line, the evaporating liquid has to absorb an increasing heat flow through the insulation and therefore has a reducing capacity for cooling the pipeline. The velocity of the liquid front therefore slows down. In the extreme case of poor insulation, the liquid front may become stationary, and there cannot be any delivery of liquid.

Zero delivery can sometimes be countered by the use of cooldown vents along the line which are closed in turn as the liquid passes them.

Zero delivery can, of course, be avoided by keeping the lines short, and having adequate insulation.

## 6.14 Pressure Surges and the Need for Ten Second Opening and Closing Times for Liquid Valves

The sudden introduction of cryogenic liquid into a warm transfer line, by the fast opening of the inlet valve, will produce a pressure surge which may lead to disaster. An initial slug of liquid will pass down the line with its front face evaporating rapidly and producing a large quantity of vapour. The pressure rise associated with this evaporation may exceed the inlet pressure and cause the unevaporated slug of liquid to flow backwards through the inlet valve. This reverse flow will be even more violent if the evaporating liquid in the hot line boils explosively by homogeneous nucleate boiling in the large temperature difference.

The reverse flow of liquid may therefore give rise to a high velocity and lead to:

- reverse liquid flow through the rotating pump, causing mechanical damage such as stripping the rotor and guide vanes,
- gate valves being slammed shut,
- line breakage and spillage.

All these have happened in the past, including a LOX transfer accident event at NASA, Cape Canaveral, USA [5].

From subsequent studies at NBS, Boulder, Colorado, it was recommended that liquid valves, whether operated hydraulically, electromagnetically or manually, should open and close over a period of 10 s or so, to avoid pressure surges and oscillations, and consequent flow reversal during liquid transfer; also that no gate valves should be used in liquid transfer lines.

It is usual for electromagnetically or hydraulically driven valves to be opened quickly in about 1 s. This is not acceptable with cryogenic liquids, and a 10 s opening and closing time should be built in.

Emergency stop and vent valves should also act slowly.

## 6.15 Care with Topping-Out

With expensive cryogenic liquids, the boil-off vapour is usually returned or collected via a vent and return line. If the vessel is over-filled when topping-out, due, say, to malfunction of the high liquid level gauge, then liquid will pass into the warm return line, where it will evaporate and create an overpressure problem.

Great care is needed to handle this problem on the spot, remembering that:

- cryogenic liquids are compressible, and will expand with decreasing pressure,
- the tank is brim full of liquid.

The overpressure needs to be very gently released, either by venting to the atmosphere if safe to do so, or by cracking open the valve to the boil-off gas compressor WITHOUT causing further liquid from the 100% full tank to pass into the return line. Any panic opening of valves to the compressor will certainly pull further liquid into the return line and accentuate the overpressure build-up.

A topping-out incident of this nature led to a 125,000 $m^3$ LNG marine tanker being grossly overpressurised—and put out of commission—due to a very expensive mistake by the loading operator, costing something in the region of $100 million. The tanker had to be emptied at the loading terminal, dry-docked and the inner tanks and insulation removed and rebuilt!

## References


1. Martinelli, R.C., Lockhart, R.W.: Proposed correlation of states for isothermal 2-phase flow in pipes. Chem. Eng. Progr. **45**(1), 39 (1949)
2. Richards, R.J., Steward, W.G., Jacobs, R.B.: Transfer of liquid hydrogen through uninsulated lines. Adv. Cryog. Eng. **5**, 103 (1960)
3. Shen, P.S., Jao, Y.W.: Pressure drop of 2-phase flows in a pipeline with longitudinal variations in heat flux. Adv. Cryog. Eng. **15**, 378 (1970)
4. Shapiro, A.H.: The dynamics and thermodynamics of compressible fluid flow. Ronald Press (1953)
5. Edeskuty, F.J., Stewart, W.F.: Safety with handling of cryogenic fluids. Plenum Press (1996)

# Chapter 7
# Design: Some Comments on the Design of Low-Loss Storage Vessels, Containers and Tanks

## 7.1 General Remarks

### *7.1.1 Three Types of Insulation*

To squeeze the subject of the design of cryogenic liquid vessels and tanks into a single chapter is unrealistic, so a few points are made on material and thermal considerations within the context of three types of insulated container employing respectively:

(a) High vacuum at $P < 1$ Pa, (with silvering, LIN cooled shield, vapour cooled shield, MLI, or multi-shielding).
(b) Low vacuum at $P \sim 10$–100 Pa, (with evacuated powders).
(c) Gas purged insulation at $P \sim 100{,}000$ Pa, just above 1 atm (perlite, PUF, or foam glass).

These points will then be further elaborated in the second part of this chapter, by looking at the thermal design of twelve widely different applications with dimensions typical of everyday use.

### *7.1.2 Heat Break Materials*

In all containers, the cryogenic liquid has to be isolated from the ambient environment by heat break materials of adequate strength and rigidity to contain the liquid under all operating conditions in a reliable manner. The heat break materials, whether to be used for necks, tank walls, pipework or supporting struts, therefore need to have high mechanical strength and low thermal conductivity, at the same time, over the whole temperature range spanned by the heat break temperature interval, e.g. between ambient and the boiling point of the cryogenic liquid.



T. D. Bostock and R. G. Scurlock, *Low-Loss Storage and Handling of Cryogenic Liquids*, International Cryogenics Monograph Series,
https://doi.org/10.1007/978-3-030-10641-6_7

Because many materials become brittle at low temperatures, the choice of heat break materials is restricted to a relatively small number of metals, mainly nickel steels, like 9% Ni steel, other nickel alloys like Invar, and some austenitic stainless steels, together with a small number of composite fibre-reinforced polymers developed specifically for making support struts [1]. Other suitable materials include certain plastic composites for the necks of small containers, and cryogenic concrete for the wall, floor and roof of large tanks [2, 3].

Polymers with either paper or fabric reinforcement are widely used as Dewar necks, components in superconducting magnets and superconductor motor generators. Polymers are commonly used where electrical resistance is required in addition to thermal resistance. Reinforced polymers used for cryogenic applications are often phenolic or epoxy based; both of which offer high electrical resistance. Particular grades of these polymers also are suitable for precision machining and as such can be also used for bushings in rotating applications where the low thermal conductivity is required.

Certain grades of fluoropolymers are also used for cryogenic applications; most commonly used are PTFE, ECTFE, FEP and PCTFE.

PEEK is used for valve seat materials and also as electrical insulator on superconducting tape, typically cost would lead one to choose another polymer option over PEEK for heat break applications.

### *7.1.3 Isothermal Containment*

While all the heat break materials can be used for containment, it is desirable to have high thermal conductivity materials for isothermal containment and as an aid towards reducing stratification, by using cheaper and lighter alternatives like aluminium and aluminium alloys. More expensive copper and copper alloys were used in the past, whilst for space applications, there are much more expensive titanium alloys for ultra-lightweight containment.

## 7.2 Trouble-Free Joints and Materials

### *7.2.1 Avoiding Joints Between Materials with Dissimilar Thermal Contractions*

A great deal of time is lost through leaks developing in joints, when they are cooled, and it is very worthwhile putting some thought into the design of joints, whether they are permanently made by welding, or are demountable to meet operational requirements.

Joints between dissimilar materials, like stainless steel and aluminium alloy, present a particular problem on account of their different thermal contraction coefficients. At low temperatures, the total differential thermal contraction between a dissimilar pair of materials may induce huge mechanical stresses, comparable to or larger than the yield stresses of the two materials, which can lead directly to leaks and mechanical failures.

For example, the thermal contraction between 300 and 77 K of stainless steel is 0.29% while that for aluminium is 0.40%. The difference of 0.11% is significant, and similar in magnitude to the yield strains for both stainless steel and aluminium. There is therefore a problem in making a mechanically reliable and leak-tight joint between the two materials.

Aluminium cannot be soldered or welded directly to stainless steel, and one way forward is the use of intermediate friction-welded joints between the two materials, up to a limited diameter.

For plastics and plastic composites, the thermal contraction between 300 and 77 K can be as large as 1.0–2.0%. However, by developing composites of epoxy resin with varying proportions of fillers such as glass fibre or zirconium silicate, the thermal contraction can be reduced to match the figure of 0.40% for aluminium and enable joints to be made between these purpose built composites and aluminium [1].

Having achieved a thermal contraction match, there still remains the joining problem. An adhesive has to be developed, such as a powder-filled epoxy, with thermal contraction properties identical to those of the pair of materials to be joined, by adjusting the powder/epoxy ratio.

We are only too familiar with failure in cryogenic systems arising from these two problems, and the consequent frustration and waste of time. Vacuum integrity of materials and joints at low temperatures are of absolutely paramount importance in systems employing vacuum insulation. If the joint problem is not properly addressed in design and manufacture, then vacuum integrity will just not be achieved, during the operational lifetime of the system.

The problems can, of course, be circumvented by using the same low thermal conductivity heat break material (such as austenitic stainless steel, 9% Ni steel or Invar) throughout, including liquid containment. Joints for low temperature use between identical alloy materials can be seam-welded as standard practice. Provided the intermediate weld-metal has the same thermal contraction, there are no local stresses at the joints from differential contraction at low temperatures; except during fast cooldowns and warm-ups.

### *7.2.2 Porosity and High Vacuum*

Vacuum integrity can still be lost if the materials are porous to atmospheric air, or to the cryogen being contained, or if the materials contain undesirable adsorbed gases such as hydrogen.

The first example of porosity one usually meets is the porous nature of Pyrex glass to helium at ambient temperature and the need to employ relatively non-porous Monax glass in all-glass liquid helium Dewars. For ultra-high vacuum systems, even Monax glass is not good enough because of its finite porosity, and helium has to be totally excluded from the vicinity.

Fortunately, Pyrex glass is non-porous to nitrogen and hydrogen, enabling it to be employed for all non-helium cryogenic systems, including the giant 3 m long, 30 cm diameter glass Dewars used at Southampton for flow visualisation and enhanced heat transfer studies.

Today, the intrinsic porosity of plastics to low MW vapour molecules, with their "large hole" molecular structures, is a major disappointment in the use of plastic composites for the vacuum enclosures of cryogenic systems.

Thin-walled austenitic stainless steel tubing can also become porous in several ways. Some of these are discussed in Sect. 7.2.3. Additionally, it must be stressed that, if the tubing is exposed to the vapour of chloride-based soft-solder fluxes, like Baker's fluid, it will surely become leaky over a surprisingly short period of time. We once happened to store a new stock of stainless steel tubing in a workshop adjacent to a soldering bay and the whole stock had to be discarded. Every cryostat built with this tubing was faulty and developed low temperature leaks, thereby wasting a great deal of time and effort.

The emergence of hydrogen as a fuel for road, air and rocket propulsion development is driving the need for lighter weight cryogenic containment, often utilising fibre reinforced polymers (FRP) and carbon fibre reinforced polymers (CFRP). There is reasonable research into FRP/CFRP as a tank material. It is fundamental to any tank that the materials used can contain the tank fluid and if vacuum insulation is to be implemented, the outgassing of the polymeric base of the FRP/CFRP must also be addressed.

Techniques to reduce outgassing and permeability are similar. Several papers [5–9] describe the development of these techniques to reduce outgassing of FRP Helium.

### *7.2.3 Porosity Problems of Austenitic Stainless Steels Transforming to Martensite*

It must be remembered that the non-brittle austenitic form of stainless steel is the high temperature equilibrium phase above about 900 °C. Below this temperature, the equilibrium phase is the martensite form, which is brittle at low temperatures. To stabilise the austenite phase against transformation to martensite, a variety of additives are included to make a range of austenitic stainless steels usable at temperatures below 900 °C. Some of these steels turn out to be generally suitable for use in cryogenic systems [4].

However, it is common experience that the austenite-martensite transformation can be induced by thermal cycling between ambient and cryogenic temperatures. The net result is that cryostats made of thin-walled austenitic stainless steel tubing, become susceptible to vacuum leaks over a period of 5–10 years. This is because the austenite—martensite transformation has taken place progressively during cooldowns and warm-ups; and while austenite is ductile at low temperatures, martensite is brittle and leaky. The austenite-martensite transformation is a one way process and irreversible.

When thin-walled austenitic stainless steel tubing starts to leak, a single leak will be followed by another, and so on. Any laborious leak test and repair routine will rapidly become a wasted effort. It must be accepted that the tubing has become porous from thermal cycling, and the only solution is to cut it out and replace it with freshly manufactured material.

By thin-walled, is meant a wall thickness of less than about 1 mm, which is generally used in laboratory cryostats and small storage vessels. When the wall thickness is greater than 1 mm, as in large vacuum insulated vessels, the development of leaks due to austenite-martensite transformation appears not to be such a problem. Possibly this is because large storage vessels are not thermally cycled very much, unlike laboratory cryostats.

It was also found at the Institute of Cryogenics, Southampton that mechanical vibration, like that in the 3000 rpm rotating cryostats, appeared to induce austenite-martensite transformation, causing the rotors to develop vacuum leaks over short periods of 1–2 years operation. Again, the only solution was to replace the stainless steel rotor necks as soon as a leak appeared, and not go through a laborious leak test and hopeful repair procedure.

It is perhaps significant that few laboratory cryostats containing thin-walled, austenitic stainless steel tubing have a lifetime greater than 10 years. The only helium cryostats which are still operational after 30 years contain Ni alloy (German silver) tubing; certainly not austenitic stainless steel.

### 7.2.4 Adsorbed Hydrogen and High Vacuum

Another problem with stainless steels is the vast amount of hydrogen which can be adsorbed during manufacture. In particular, the process of nitriding so as to produce a shiny finish, of "bright" steel, by a high temperature reaction in an ammonia atmosphere, leads to high amounts of adsorbed hydrogen.

If this hydrogen is not removed before incorporating the shiny plate or tubing into a storage vessel or cryostat, then the vacuum insulation will tend to be impaired over a relatively short time by the continuous desorption of the hydrogen into the vacuum space. The solution is to vacuum bake the "bright" steel at 300 °C for 24 h, to get rid of all the adsorbed hydrogen, before use in manufacturing cryogenic vessels.

### 7.2.5 Frost-Proof Cryogenic Concrete

Free water in concrete will freeze to ice on cooling to −10 to −60 °C, with significant expansion leading to cracking and spalling. On the other hand, the chemically bound water does not contribute to this expansion.

The simple solution to making frost-proof concrete is to remove all of the free water after curing of the fresh concrete, and preventing rewetting from rain and moist air with a water repellant or sealant coating. In addition, of course, the aggregate must have the same expansion coefficient as the cement paste [3]. The concrete is then frost-proof and capable of reliable usage down to liquid helium temperatures, without breaking up or reduction of mechanical strength.

### 7.2.6 Hydrogen Embrittlement

Hydrogen embrittlement occurs when molecular hydrogen diffuses into the structure of the metal. The hydrogen forms hydrides and hydrogen induced martensite [6] resulting in a loss of ductility, potentially crack initiation and failure below the yield stress.

Hydrogen can be introduced to metal through manufacturing. Welding, electroplating, phosphating and pickling introduce hydrogen into the metallic structure during the manufacturing process. This can be mitigated by heat treatment of the material after manufacturing.

Hydrogen can be introduced into materials from the environment and of course metalwork for hydrogen service will naturally see more hydrogen than that used in other environments. Embrittlement effects are reduced at low temperature; however all cryogenic vessels will be at ambient temperatures at times during their service life.

Hydrogen introduced to stainless steels will diffuse into the material and react with the alloyed carbon. This leads to a reduction in mechanical strength and ductility which can result in cracking and failure. At elevated temperatures, hydrogen will diffuse into the steel structure and accumulate at grain boundaries. At ambient temperatures the hydrogen will accumulate at inclusions, with similar resultant effects.

Welding of stainless steel leads to the formation of delta ferrite within the base material. Delta ferrite at a certain level increases mechanical strength, and protects against stress corrosion cracking. Delta ferrite is similar to martensite and both form in similar areas of the lattice, and whilst it may be assumed that delta ferrite may promote martensite growth, Michler concludes that delta ferrite does not lead to the formation of martensite [8].

## 7.3 Thermal Considerations

### *7.3.1 Choice of Boil-off Rate*

The starting point is the choice of boil-off rate, which determines the design heat flux per unit area into the cryogenic system. The choice depends upon a number of considerations such as:

(1) The purpose of the system, whether to cool an instrument, a superconducting magnet, or an experimental cryostat, or to store a cryogenic liquid for use as a coolant or for subsequent re-gasification.
(2) The temperature of the cryogen(s) to be used, which will determine the type of insulation needed.
(3) The required mobility of the system, whether it is a tank on a space rocket, a sea-going tanker, a road or rail tanker, or a static storage Dewar, vessel or tank.
(4) The mode of operation of the cryogenic system, whether continuous, or intermittent. The latter mode will have cooldown requirements, dictating either the size of refrigeration plant or the peak demands for cryogenic liquid.
(5) The required operational lifetime of a single filling of cryogen, or possibly the inverse, namely the percentage boil-off loss of cryogen per day.
(6) The volume of liquid being stored and hence the scale of the storage system to be designed.
(7) Whether there is a requirement for cryogen-free operation using refrigeration plant or cryocoolers, when the design heat flux will be determined by the refrigeration system to be used.

This enormous range of considerations arise because there are so many applications of cryogenics today [5].

### *7.3.2 Some Practical Applications*

To illustrate the scale of cryogenic use today, the thermal design of twelve totally different, but typical, applications are considered briefly in Sect. 7.4, to end this chapter.

The scales concerning the volumes of "cold" being used extend from 10 mm diameter for NMR/FTIR sample holders, to 75 and 100 m diameter and 50 m high for LNG and LPG static storage tanks.

The starting point in all designs is the boil-off rate, or percentage loss-rate, so it is useful to tabulate (Table 7.1) what is acceptable today, for the twelve applications, before looking briefly at the designs.

**Table 7.1** Boil-off performance of applications

| | Liquid volume | Duration of fill | Daily boil-off% |
|---|---|---|---|
| 1. NMR/FTIR sample cell | 4 ml | 10 min | 14,400 |
| 2. LHe lab cryostat | 2 l | 120 h | 20 |
| 3. LHe storage dewar | 500 l | 100 days | 1.0 |
| 4. MRI cryostat | 400 l | 100 days | 1.0 |
| 5. Static LHe storage | 12,000 l | 333 days | 0.3 |
| 6. LHe space probe | 4.000 l | 1000 days | 0.1 |
| 7. LOX rail tank | 48 $m^3$ | 200 days | 0.5 |
| 8. LIN tank, dustbin | 1700 $m^3$ | 200 days | 0.5 |
| 9. LIN tank, cluster | 1100 $m^3$ | 333 days | 0.3 |
| 10. LNG tanker | 125,000 $m^3$ | 1000 days | 0.1 |
| 11. Static LNG tank | 220,000 $m^3$ | 2000 days | 0.05 |
| 12. Static LPG tank | 390,000 $m^3$ | 10,000 days | 0.01 |

### 7.3.3 Heat Fluxes Through Insulations in Practical Applications

The specified boil-off rates enable the allowed heat fluxes through the insulation to be calculated (Table 7.2). The next stage in the thermal design is the selection of insulations to achieve these heat fluxes. These are outlined in Sect. 7.4, for each of the twelve applications.

It should be noted that the quoted figures of boil-off rates and hence allowed heat fluxes are probably accurate to within a factor of two. They do however give an

**Table 7.2** Average heat fluxes of applications

| | Volume | Wetted area | Heat in-flow | Av. heat flux |
|---|---|---|---|---|
| 1. NMR sample cell | 4.7 $cm^3$ | 20 $cm^2$ | 1070 mW | 535 $W/m^2$ |
| 2. LHe cryostat | 0.004 $m^3$ | 0.165 $m^2$ | 11.6 mW | 70 $mW/m^2$ |
| 3. LHe storage | 0.52 $m^3$ | 3.14 $m^2$ | 145 mW | 46 $mW/m^2$ |
| 4. MRI cryostat | 2.35 $m^3$ | 14.2 $m^2$ | 116 mW | 8.2 $mW/m^2$ |
| 5. Static LHe storage | 12.6 $m^3$ | 31.4 $m^2$ | 1.1 W | 35 $mW/m^2$ |
| 6. LHe space probe | 4.18 $m^3$ | 12.6 $m^2$ | 116 mW | 9.2 $mW/m^2$ |
| 7. LOX rail tank | 48 $m^3$ | 75 $m^2$ | 680 W | 9.0 $W/m^2$ |
| 8. LIN tank, dustbin | 1720 $m^3$ | 796 $m^2$ | 15.9 kW | 20.0 $W/m^2$ |
| 9. LIN tank, cluster | 1142 $m^3$ | 1320 $m^2$ | 10.5 kW | 8.0 $W/m^2$ |
| 10. LNG tanker | 5 × 25,400 $m^3$ | 5 × 4200 $m^2$ | 5 × 63.6 kW | 15.2 $W/m^2$ |
| 11. Static LNG tank | 220,000 $m^3$ | 16,200 $m^2$ | 250 kW | 15.4 $W/m^2$ |
| 12. Static LPG tank | 390,000 $m^3$ | 23,500 $m^2$ | 112 kW | 4.8 $W/m^2$ |

insight into the scale of cryogenic liquid applications in general use today and the effectiveness of the insulations in use.

One interesting feature is the similarity in heat fluxes for the LHe systems, apart from the low value for the MRI cryostat, and for the LIN/LOX/LNG systems. Despite the enormous size of the LPG tank, the average heat flux is about one third that of the LNG tank.

## 7.4 Thermal Design of 12 Typical Cryogenic Liquid Applications

### *7.4.1 LIN Cooled Sample Holder, 10 mm Diameter, 60 mm Long*

Such a sample holder, 10 mm diameter, 60 mm long, is used to keep a sample at 77 K for 10 min using a single fill of LIN to enable a set of instrument readings, such as NMR or FTIR readings, to be taken on a LIN-cooled sample. An unsilvered, all glass, double-walled, evacuated, miniature vessel is envisaged as the sample holder. No silvering is allowed for the measurements.

The volume of LIN to last 10 min absorbs, via its boil-off, a heat in-flow of 1070 mW, corresponding to a heat flux into the LIN of 535 W/m$^2$.

Now, the experimentally quoted heat flux through an unsilvered, evacuated, double vessel is 500 W/m$^2$, which is about 7% less than the boil-off heat flux. The LIN will therefore last longer than 10 min, and the design just meets the sample cooling requirement.

### *7.4.2 Laboratory LHe Cryostat with Isothermal Volume, 100 mm Diameter, 500 mm Long*

This is the standard size of liquid helium cryostat, used by many research workers, and will normally incorporate a LIN cooled radiation shield surrounding the LHe bath.

Assuming that half the isothermal volume is taken up by the working system, magnet, etc., the volume of LHe is 2000 ml. Allowing five days, or a full working week of operation after filling with LHe, the design boil-off rate is 400 ml/day. This boil-off corresponds to a heat in-flow of 11.6 mW, and an average heat flux through the insulation into the isothermal volume of 70 mW/m$^2$.

Now, the radiation heat flux from 77 to 4.2 K, for aluminium/copper with e = 0.02, is 22 mW/m$^2$, or 3.3 mW into the cryostat. This leaves 8.3 mW for all other sources of heat in-flow.

Using the performance diagram. Figure 4.5a, to obtain the neck aspect ratio for zero neck conduction into the LHe, ($W_0 = 0$), L/A for a total heat flow of 11.6 mW is about 4000 $m^{-1}$, or 40 $cm^{-1}$. This latter figure provides the design guide for the neck geometry between 4.2 and 77 K. For a neck wall-thickness of 0.5 mm (0.020 in.), A is $1.5 \times 10^{-4}$ $m^2$ from which L (4.2–77 K) is 0.60 m.

Horizontal vapour-cooled radiation shields, at this height of neck (0.60 m) above the top of the isothermal volume, will ensure that the radiation heat flow down the neck into the LHe is negligible.

While MLI around the LIN cooled shield is beneficial, there is no need to have any MLI between the shield and the inner LHe vessel.

### *7.4.3 500 Litre LHe Laboratory Storage Dewar*

The storage Dewar envisaged consists of a 1 m diameter sphere with a vertical access neck for liquid transfer in and out. The storage volume is 0.523 $m^3$, so that for a design figure for the boil-off of 1% per day, or 5 l/day, the allowable heat in-flow is 145 mW, corresponding to an average heat flux through the insulation space of 46 mW/$m^2$.

The radiation heat flux from 77 K for e = 0.02 is 22 mW/$m^2$, or 66 mW for the Dewar, leaving 79 mW for other sources, transfer lines, bump stops, etc.

Using the performance diagram, Fig. 4.5a, for $W_0 = 0$, L/A (77–4 K) for a total heat flow of 145 mW is about 300 $m^{-1}$.

For a 10 cm diameter neck and neck wall of 1.0 mm (0.040 in.), A is $3.14 \times 10^{-4}$ $m^2$ from which L is 0.094 m. This neck length is surprisingly short.

For multi-shielding, instead of a LIN cooled radiation shield, the design of multi-shields requires the innermost one to be as cold as 77 K, but not any colder, with no MLI inside.

### *7.4.4 MRI Cryostat Without, and with, Cryocooler*

Assuming the inner cryostat is 2 m larger diameter, 1 m smaller diameter, the volume is 2.35 $m^3$, which includes the magnet and LHe coolant, while the cold wetted surface is 14 $m^2$.

Assuming a total of 400 l of LHe and a 1% boil-off per day of 4 l/day, the total heat flow is 116 mW, corresponding to a surface heat flux of 8.2 mW/$m^2$.

The radiation heat flux for a LIN cooled aluminium shield, e = 0.02, is 22 mW/$m^2$, so LIN cooling is not enough. With multi-shielding and an inner shield at 50 K, the radiation heat flux is 4 mW/$m^2$, or a total heat flow of 56 mW, which is about one half the total allowable heat flow and is therefore acceptable.

If neck and support heat flows can be kept below 60 mW, then a 1% daily loss rate and a hold-time of 100 days can be achieved. Eddy current heating during MRI operation will increase the boil-off rate above the daily 1% figure.

With a 2-stage 4 K cryocooler to recondense the LHe boil-off vapour and cool a radiation heat shield at 50–60 K, a LHe refill every 100 days can be avoided.

### 7.4.5 12,600 Litre Static LHe Storage Vessel, 2 m Diameter, 4 m High

The vessel consists of a cylindrical inner LHe container with a vertical access neck for liquid transfer in and out. The storage volume is 12.6 $m^3$, and for a boil-off of 0.3% per day, or 38 l/day, the allowable heat inflow is 1.1 W, corresponding to an average heat flux through the insulation space of 35 $mW/m^2$.

The radiation heat flux from 77 K for e = 0.02 is 22 $mW/m^2$, or 690 mW for the total vessel, and allows about 410 mW for other sources of heat inflow, transfer lines, bump stops, etc., which is probably not enough.

The radiation heat flux from 60 K is 8 $mW/m^2$, providing a total of 250 mW, allowing 850 mW for the other sources, which is probably better.

Using the performance diagram. Figure 4.5a, for $W_0 = 0$, L/A for a total heat in-flow of 1.1 W is about 0.4 $cm^{-1}$. For a 10 cm diameter neck and wall thickness of 2.5 mm (0.100 in.), A is 7.8 $cm^2$, for which L is only 3.2 cm between 4 and 77 K.

This is an indication of the minimum neck length needed, to reduce the conducted heat flow down the neck-wall to zero by vapour cooling.

For multi-shielding, the inner-most shield should be no colder than 60 K, with no MLI inside it. The length of the neck is determined almost entirely by the number and spacing of the thermal contact rings to the multi-shields.

### 7.4.6 4000 Litre LHe Space Probe, 2 m Diameter, 3 Year Hold Time

Consider a 2 m diameter spherical LHe container, with a hold time of three years, or a daily boil-off loss rate of 0.10% in round figures.

For a boil-off of 0.1% per day, or 4 l/day, the allowable heat inflow is 116 mW, corresponding to an average heat flux through the insulation space of 9.2 $mW/m^2$.

Multi-shielding, with the inner aluminium shield, e = 0.02 at 60 K, yields a radiation heat flux of 8 $mW/m^2$, or 100 mW for the vessel, allowing 16 mW for the other heat inflow sources. This is too tight, using the working criterion that radiation should not contribute more than 50% of the total heat inflow.

For an inner shield at 50 K, the radiation heat flux is 4 $mW/m^2$, or 50 mW total, allowing 66 mW for the other sources. This is much better.

The neck geometry of a space probe is somewhat indeterminate because the zero gravity environment will prevent convection and neck cooling, as we know it under 1 g. In which case, the design and spacing of the thermal contact rings on the vapour vent lines linking to the multi-shields will need careful attention and experimental development.

### 7.4.7 LOX Rail Tank or VIT Vessel, 3 m Diameter, 8 m Long, 48 m$^3$ Volume

Consider an 8 m long, 3 m diameter, horizontal, cylindrical tank with a required boil-off of 0.5% per day. For this boil-off, the 'A' heat in-flow is 680 W, corresponding to an average heat flux through the insulation of 9.0 W/m$^2$.

Using perlite evacuated powder insulation, with a quoted k-value of 1.0 mW/mK, the average heat flux is 4.0 W/m$^2$ for 50 mm thickness of insulation, or 300 W for the tank, allowing 380 W for the other heat inflow sources.

Since the tank will require substantial supports between inner and outer vessels, to withstand acceleration forces during rail journeys, shunting, etc., 380 W may not be enough, and a thicker filling of evacuated powder would be needed. The overall maximum dimensions, diameter and length, are limited for rail tanks, and thicker insulation means reducing the volume of LOX to be carried. The daily boil-off of 0.5% is therefore unlikely to be acceptable.

Alternatively, using MLI with a k-value of 50 μW/mK, the total heat flux can be reduced to 1.0 W/m$^2$ with only 10 mm (20 layers) of MLI. This would provide a total heat inflow of 75 W, allowing 600 W for other heat sources to meet the daily boil-off of 0.5%.

The use of one vapour cooled shield within the MLI insulation, to convert some of the 'A' heat in-flow into 'B' heat in-flow, would lower the heat flux through the MLI by a factor of, perhaps, 2–40 W—which is not worthwhile.

Such a shield with evacuated powder insulation would probably lower the heat flux by a factor of two, also, but would not be practicable mechanically.

### 7.4.8 Static LIN/LOX Tank, 13 m Diameter, 13 m High, 1720 m$^3$: Dustbin Configuration

This tank can be in two configurations, either

(a) a dustbin-type, double-walled, cylindrical tank, with an inner vessel 13 m diameter and 13 m high, or
(b) a multi-cylinder "cluster tank" composed of, say, seven in number 4 m diameter 13 m high, inner vessels inside an outer vessel 14 m in diameter.

In both cases, the insulation is Perlite powder with a nitrogen gas purge at 1 bar. For the dustbin configuration, the volume of liquid stored is 1720 $m^3$ and the required daily boil-off of 0.5% gives an 'A' heat in-flow of 15.9 kW corresponding to an average heat flux through the insulation of 20.0 $W/m^2$.

For perlite powder with nitrogen gas purge, the quoted k-value is 30 mW/mK, and for a thickness of 0.5 m, the average heat flux is 12.0 $W/m^2$ with a total wall heat inflow of 6.4 kW. However, the flat base requires a load bearing insulation such as foam glass bricks, with a k-value of 52 mW/mK, yielding a heat flux of 16.8 $W/m^2$ for a thickness of 0.7 m and a floor heat inflow of 2.2 kW.

The total estimated heat flow through tank wall and floor is then 10.1 kW, allowing 5.8 kW for reverse convection, pipework, and radiation from the tank roof. Thus it appears that the 0.5% boil-off rate per day can be achieved.

### *7.4.9 Static LIN/LA/LOX Tank, 13 m Diameter, 13 m High, 1142 $m^3$: Cluster Configuration*

For the cluster tank, seven 4 m diameter, 13 m long cylinders are mounted vertically and separated from each other and from the outer vessel by a minimum of 0.5 m of nitrogen gas purged, perlite powder insulation.

The total volume of stored liquid is 1142 $m^3$ and a lower daily boil-off figure of, say, 0.3% is required, corresponding to an 'A' heat in-flow of 10.5 kW or an average heat flux of 8.0 $W/m^2$. For a minimum thickness of 0.5 m insulation, the total heat flow is estimated to be 4.8 kW, and the required boil-off can be realised.

The advantages of the cluster tank configuration are:

(1) Lower construction costs since the 4 m diameter inner tanks of, for example, 9% nickel steel, can be factory made and transported by road to the site, (Both inner and outer dustbin tanks have to be constructed on site).
(2) A good boil-off performance compared with a dustbin tank,
(3) A mix of liquids can be conveniently stored in the separate vessels within the cluster including LIN, LA and LOX.

A disadvantage is the reduced liquid volume stored (only 66% of the dustbin tank, Sect. 7.4.8), which is offset by the lower boil-off loss rates.

### *7.4.10 Sea Tanker for 125,000 $m^3$ LNG*

There are two basic types of construction:

(a) A number of integral, membrane, prismatic-shaped tanks within the ship's hull. The membrane tanks all have different shapes so as to fit within the structure of the ship, the tanks usually being constructed of low thermal expansion coefficient Invar.

(b) A number of spherical tanks supported by cylindrical skirts as heat breaks and extending down from the equator of each tank to the ship's structure.

Consider the spherical tank design, which is simpler in structure, and the thermal performance of a single sphere. A standard size 125,000 m$^3$ LNG tanker carries five 25,000 m$^3$ spheres, each 36.5 m in diameter, beneath its insulation. A daily boil-off rate of 0.10% corresponds to a heat inflow to one sphere of 63.6 kW and a heat flux of 15.2 W/m$^2$.

For polyurethane foam PUF insulation, with a k-value of 30 mW/mK, 1 m thickness of insulation will yield a heat flux of 6.0 W/m$^2$, or a heat inflow of 25.2 kW to the tank, allowing 38.6 kW for other sources of heat inflow. The other major heat source is the equatorial support ring and cylindrical skirt.

One complication is the need for a secondary, liquid-tight, membrane within the insulation to prevent any leakage of LNG reaching the carbon steel structure of the ship. This safety requirement is mandatory, and is more easily met with spherical tanks than with integral membrane tanks.

The enormous thermal contraction, associated with the large dimensions of the ship tanks, presents a particular problem which can be met largely by extensive use of Invar.

### *7.4.11 Static LNG Tank, 75 m Diameter, 50 m High, 220,000 m$^3$ Volume*

With a liquid capacity of 220,000 m$^3$, this LNG tank has an inside diameter of 75 m and a height of 50 m. The floor area is 4415 m$^2$, with a wall area of 11,775 m$^2$.

For this very large tank, a double-walled, all-metal, design would be very expensive to build on site. Instead, the inner tank is made of 9% nickel steel, welded in sections on site. A secondary containment wall of prestressed concrete, utilising cryogenic reinforcing bar, rising to the top of the primary tank wall will be necessary to contain any leaks of LNG from the primary tank. The space between the inner and outer containment walls is typically insulated with loose-fill expanded perlite. The floor is insulated with cellular glass bricks.

Instead, the construction material will be pre-stressed cryogenic concrete, together with a thin, membrane, liquid-tight liner of stainless steel or Invar inside. The inner, primary, concrete tank will be insulated on the outside by PUF round the walls and cellular glass bricks below the floor.

To prevent frost heave of the underlying ground, a heater mat will have to be installed below the floor insulation. Alternatively, the whole primary tank structure could be supported on piles clear of the ground and the bottom of the tank heated by air flowing under the tank.

The rationales for heat mats, and piling will be influenced by the tanks ambient conditions, geotechnical investigation and seismic zone.

For a boil-off of 0.10% per day, the allowable heat inflow is 204 kW. For a thickness of 1 m of PUF insulation, k-value 30 mW/mK, the heat flux is 6.0 W/m$^2$, and the corresponding wall heat inflow is 70 kW. For a thickness of 1 m of cellular glass, with a k-value of 45 mW/mK, the heat flux is 9.0 W/m$^2$ and the floor heat flow is 40 kW, which equals the power input to the under-floor heater mat.

The total heat flow through the insulation is then 110 kW, allowing about 100 kW from other sources such as reverse convection in the vapour core, pipework, and the suspended-deck roof insulation, the latter providing a radiation heat in-flow of about 23 kW, for a deck temperature of 200 K with e = 0.1.

Note that the under-floor heater power of 40 kW could be reduced by employing a thicker layer of cellular glass, load bearing insulation, say 2 m thickness instead of 1 m.

### *7.4.12 LPG Tank, 100 m Diameter, 50 m High, 390,000 m$^3$ Volume*

This very large tank is typical of a number of LPG tanks in operation, and is chosen as the final example of the containment of cryogenic liquids in the world today. The size of the whole structure is difficult to comprehend, but it is as large as a modern sports stadium seating 40,000 people.

The inner, primary tank has an inner diameter of 100 m, a height of 50 m, a liquid volume of 390,000 m$^3$, a floor area of 7850 m$^2$ and a wetted-wall surface area of 15,700 m$^2$.

The construction of the primary tank will almost certainly be of pre-stressed cryogenic concrete, with an Invar, liquid-tight, inner liner. The roof is likely to be constructed of radial pre-stressed concrete beams supported at one end by the top of the wall and at the other by a central pillar built up from the floor.

A secondary concrete wall and earth bund surrounds the primary structure, to provide secondary liquid containment in the event of a spillage from the primary tank.

The floor insulation has to be load bearing and is therefore built of cellular glass or perlite concrete bricks. Frost heave of the underlying ground is a major problem and either heater mats are incorporated below the floor insulation, or the whole tank is supported on piles clear of the ground.

For a daily boil-off of 0.01% of liquid propane at 231 K (−42 °C), the allowable heat in-flow is 112 kW.

For a thickness of 1 m of foam glass, with a k-value of 50 mW/mK, the heat flux through the floor is 3.0 W/m$^2$, corresponding to a heat in-flow of 23.5 kW. For 1 m of PUF insulation, with a k-value of 30 mW/mK, applied to the wall and roof, the heat flux is 1.8 W/m$^2$, corresponding to a heat inflow of 42 kW. The total heat inflow through roof, walls and floor is then 65.5 kW. This leaves a heat flow of 46.5 kW for other sources such as reverse convective core flow and pipework.

The heat load of heater mats, installed below the tank floor to prevent frost heave, is 23.5 kW; the same as the heat in-flow through the floor into the LPG.

## 7.5 Summary of Thermal Design of Cryogenic Liquid Vessels

1. Determine the boil-off rate and holding time of the contained liquid.
2. Identify all heat inflows and reduce them to the same magnitude using appropriate insulation.
3. Design system to convert 'A' heat in-flows into 'B' heat in-flows.
4. Note that reverse convection in the vapour column cores will probably be the major source of 'A' heat in-flow.
5. For multi-shielding of most LHe vessels, the inner shield should be no colder than 60 K, with no MLI inside the shield.

## References, Specific

1. Hartwig, G., Evans, D.: Non-Metallic Materials and Composites. Plenum Press (1982)
2. Turner, F.H.: Concrete and Cryogenics. Viewpoint Publications (1979)
3. Scurlock, R.G., Mohd Yusof, K.B.: Cryogenic and frostproof concrete made from conventional aggregates and Portland cement. Adv. Cryog. Eng. **45**, 1779 (2000)
4. Wigley, D.A.: Mechanical Properties of Materials at Low Temperatures. Plenum Press (1971)
5. Kalia, S., Fu, S.: Polymers at Cryogenic Temperatures. Springer, Berlin, Heidelberg (2013)
6. Bechel, V.T., Negilski, M., James, J.: Limiting the permeability of composites for cryogenic applications. Compos. Sci. Technol. **66** (2006)
7. Disdiera, S., Reya, J.M., Paillera, P., Bunsellb, A.R.: Helium permeation in composites materials for cryogenic application. Cryogenics **38** (1998)
8. Lyako, V.Y., Fedorchenko, A.V., Kivurenko, O.B., Shnyrkov, V.I.: FRP Dewar for measurements in high pulsed magnetic fields. Cryogenics **49** (2009)
9. Xu, L., Wang, R., Lu, X.: The vacuum characteristics of FRP liquid helium Dewars. ICEC (1998)

## References, General (in Reverse Order of Publication)

10. Schultheiß, D.: Permeation barrier for lightweight liquid hydrogen tanks. OPUS Augsburg (2007)
11. Michler, T.: Toughness and hydrogen compatibility of austenitic stainless steel welds at cryogenic temperatures. Hydrog. Energy **32** (2007)
12. Robinson, S.L., Somerday, B.P., Moody Sandia, N.R.: Hydrogen Embrittlement of Stainless Steels. National Laboratories, Livermore, CA, USA

13. Pan, C., Su, Y.J., Chu, W.Y., Li, Z.B., Liang, D.T., Qiao, L.J.: Hydrogen embrittlement of weld metal of austenitic stainless steels. Corros. Sci. **44** (2002)
14. Flynn, T.M.: Cryogenic Engineering, 2nd edn. Taylor and Francis (2004)
15. Cryogenic Fluids Databook. British Cryogenics Council (2002)
16. GASPAK Cryodata Inc. e-mail: sales@htss.com
17. White, G.K.: Experimental Techniques in Low Temperature Physics, 4th edn. Oxford University Press (2002)
18. Isalski, W.H.: Separation of Gases. Oxford University Press (1989)
19. Timmerhaus, K.D., Flynn, T.M.: Cryogenic Process Engineering. Plenum Press (1989)
20. Hands, B.A.: Cryogenic Engineering. Academic Press (1986)
21. Van Sciver, S.W.: Helium Cryogenics. Plenum Press (1986)
22. Barron, R.F.: Cryogenic Systems 2. Oxford University Press (1985)
23. Williams A.F., & Lorn W.L.: Liquefied Petroleum Gases. Ellis Horwood (1982)
24. Arkharov, A., Marfenina, I., Mikulin, E.: Theory and Design of Cryogenic Systems. Mir Publishers, Moscow (1981)
25. Frost, W.: Heat Transfer at Low Temperatures. Plenum Press (1975)
26. Haselden, G.G.: Cryogenic Fundamentals. Academic Press (1971)
27. Thermodynamic Properties of Refrigerants. ASHRAE (1964)
28. Kropschot, R.H., Birmingham, B.W., Mann, D.B.: Technology of Liquid Helium. NBS Monograph **111** (1968)
29. Scott, R.B.: Cryogenic Engineering. Van Nostrand (1959), 6th reprint (1967)
30. Scott, R.B., Denton, W.H., Nicholls, C.M.: Technology and Uses of Liquid Hydrogen. Pergamon Press (1964)
31. Din, F.: Thermodynamic Functions of Gases, vol. 1–4. Butterworths (1962)
32. Hoare, F.E., Jackson, L.C., Kurti, N.: Experimental Cryophysics. Butterworths (1961)
33. Ruhemann, M.: The Separation of Gases, 2nd edn. Oxford University Press (1949)

References (Actual, in Reverse Order of Publication)


[illegible] Su, V.J., Chu, W.Y., Hsia, B., [illegible], D.T., Qian, [illegible] [illegible] [illegible] (2011) [illegible]

[illegible]

# Chapter 8
# Safe Handling and Storage of Cryogenic Liquids

## 8.1 General Remarks

The need to conduct the handling and storage of cryogenic liquids in a safe and responsible manner is obvious for moral, environmental and economic reasons. A good advisory first point of reference is "The Cryogenics Safety Manual: A guide to good practice" published by the British Cryogenics Council [1].

In addition, we are, today, all subject to responsibility for safety. Specific safety guidelines are laid down by national and international legislation as in the US and Europe. In the UK, the Health and Safety Executive (HSE) is the working body for guidelines on safety procedures, accident reporting and analysis, and for legal action if necessary.

*The Cryogenics Safety Manual* is a useful guide to good practice for all operators and users handling cryogenic fluids. The Manual was first published in 1970 and it has since been revised in 1982, 1991, 1998 and 2018.

Other references on safety include the volume by Zabetakis, on *Safety with Cryogenic Fluids*, published in 1967 [2], and the volume by Edeskuty and Stewart on *Safety in the Handling of Cryogenic Fluids*, published in 1996 [3].

This final, advisory, chapter does not intend to cover all the safety material in the Cryogenics Safety Manual. It does, however, contain a useful description of the building and operation of a cryogenic safety laboratory. It also contains many advisory points, arising from earlier chapters of this book, which are additional to those made in the *Cryogenics Safety Manual*, and includes some personal experiences of handling many cryogenic liquids over some 60 years.



T. D. Bostock and R. G. Scurlock, *Low-Loss Storage and Handling of Cryogenic Liquids*, International Cryogenics Monograph Series,
https://doi.org/10.1007/978-3-030-10641-6_8

## 8.2 Health Concerns

### *8.2.1 Cold Burns*

Human tissue cooled below about −10 °C will suffer necrosis or frost bite or cold burns. It is therefore sensible to ensure that, when handling cryogenic liquids, the right clothes are worn so that splashes and spills do not result in cold burns. These should include loose gloves, long sleeves, trousers without turn-ups, shoes/boots—not sandals—eye protectors and so on. First Aid for a cold burn is the application of cold water (*not* hot water) as fast as possible, since seconds wasted will make all the difference between serious injury and just a stinging sore patch.

### *8.2.2 Asphyxia and Anoxia*

Any spillage of cryogenic liquid, or cooldown process, will result in the generation of large volumes of cold, dense vapour. Cold vapour will firstly flow downwards and displace atmospheric air upwards, and secondly mix with air, diluting the oxygen content and forming a dense mist of condensed water particles. The mist, and the clear space below it down to ground level, is deficient in oxygen and is therefore dangerous to breathe. If the oxygen content is less than 6% then sudden and acute asphyxia will result in instant unconsciousness, without any physiological warning. As described in the BCC Cryogenic Safety Manual, "The victim will fall to the ground as if struck down by a blow to the head and will die in a few minutes unless immediate remedial action is taken".

Fatal accidents happen every year from sudden asphyxia, generally with the use of LIN. If a major spillage occurs, evacuate the area immediately, and keep away from the mist of cold vapour.

Gradual asphyxia, or anoxia, also with fatal consequences, can arise from progressive lowering of the oxygen content by poor venting procedures over a period of time. Asphyxia can be avoided by the use of firstly, common sense and secondly, personal gas monitors, which produce an alarm signal if the oxygen content drops below 20.9% (say to 19%). The important thing is to be aware of the deadly danger of asphyxia, with no physiological warning of oxygen deficiency (breathlessness is only brought on by the build up of carbon dioxide) and the danger of complacency about the hazards leading to asphyxia.

## 8.3 Equipment Failure

### 8.3.1 *Materials*

Certain materials are not suitable for use in cryogenic systems because of the danger of brittle failure. Such materials include, for example, carbon steels and plastics and they must not be used.

Generally, metal fatigue limits are higher at low temperatures so that fatigue failure is less likely. However, brittle failure of pipes and tanks can lead to spillages and serious consequences.

### 8.3.2 *Overpressure*

Overpressure in a pipe line, following shutdown after a transfer, must be prevented by protecting each valved-off section with its own pressure-relief valve and adequately sized vent. Vacuum spaces are a particular problem, because a leak from the inner liquid-containing vessel can lead to overpressure in the vacuum space and subsequent collapse of the inner vessel. All vacuum spaces should be protected by blow-off discs to protect against overpressure.

### 8.3.3 *Spillage Containment*

A major spillage of cryogenic liquid can lead to serious and tragic consequences, so the concept of secondary containment of the spill for a short period needs to be applied. This can take the form of:

(1) A berm, bund or earth bank to prevent a major spill spreading into the local environment.
(2) A secondary, liquid containing, wall as for LNG tanks. It is standard to use cryogenic rebar mesh, within the secondary wall of large, concrete, cryogenic tanks, to ensure the structure is capable of containing a spill.
(3) A secondary barrier or membrane within the insulation space, as in LNG ship tankers, to prevent contact between LNG and the carbon steel structure of the ships. Additionally, it has been found that the secondary membrane also provides considerable anti-spill protection during ship-ship collisions and ship-shore groundings.
(4) Adequate integrity of the outer containment structure of the applied insulation.

The nightmare scenarios of spillage accidents generally arise with single skinned containers such as LPG pressurised road and rail tankers and the older crude petroleum tankers.

Spillage due to transfer line breakages can be prevented by the installation of automatic shut-off valves inside storage tanks and vessels.

### *8.3.4 Fire and Explosion*

The conventional description of fire is the coincidence between:

(1) combustible
(2) oxidant and
(3) ignition source.

Fire protection is the rigorous separation of these three items; a combination of any two represents a potential fire hazard.

Among the cryogenic liquids, hydrogen is a particular combustible which will be discussed below (see 8.5.3). Hydrocarbons, such as LNG or LPG, are generally flammable in air over limited composition ranges. For example, methane is only flammable between compositions of 5 and 15% by volume in air. The density of methane gas is about one half that of air, so the vapour from LNG rises rapidly, mixes by convection with the air and quickly dilutes to below the lower flammability limit of 5% while dispersing upwards and away from ignition sources.

Heavier hydrocarbons, like LPG, which evaporate to produce a vapour more dense than air present a more serious fire hazard. The vapour stratifies and

(a) does not mix convectively with less dense air, and thereby tends not to dilute to below the lower flammability limit, and
(b) does not disperse by rising upwards.

Instead, the dense vapour cloud remains close to the ground, flowing downhill via topographical features, as a flammable mixture. There is also a danger that the vapour cloud can accumulate in low points and confined spaces such as pits, trenches or cellars.

A vapour fire in an enclosed space, or in the open air, can run up into an explosion.

Fires associated with LPG storage tanks can be extremely dangerous. If gas released from a vent valve produces a large flame, the heat from the flame will cause the tank pressure to rise, as well as weakening the tank wall. Subsequent tank failure may be accompanied by a Boiling Liquid and Expanding Vapour Explosion (BLEVE). Fireballs several hundred meters in diameter are known to have been created by BLEVEs, some with many casualties suffered as a result of burns.

Apart from air, LOX is a powerful oxidant and a major fire hazard. Fires fed by LOX are so hot they cannot be extinguished by foam techniques. Materials not normally flammable in air (such as certain plastic materials) may burn fiercely in LOX. Water is the most effective medium for controlling a LOX-fed fire.

Apart from ignition sources created by high or low voltage electrical faults, it should be remembered that electrostatic charges can also create ignition sources in the form of sparks. All cryogenic liquids have exceedingly low electrical conductivities and their movement from vessels and tanks through pipelines or hoses, which are not adequately earthed, can induce charge separation and high voltages.

Atmospheric oxygen will condense on any pipe work with a surface temperature lower than 90 K, it is common to see oxygen condensing on poorly insulated LIN transfer lines and valves. Caution must be taken for this LOX not to collect nor saturate any fibrous material, including insulation, clothes or carpet. The preferred solution is to better insulate the lines/valves in use to eliminate or reduce any condensation.

## 8.4 Liquid Management Problems

### *8.4.1 Overfilling of Vessel or Tank*

Cryogenic liquids have finite volume coefficients of compressibility, and they expand significantly with increasing temperature along the saturation line (see Table 5.1) or with isenthalpic and isentropic decreases in pressure. This behaviour is totally unlike that of water, which has almost zero compressibility; thus water is a useful liquid, for example, for hydraulic power transmission and pressure testing of vessels.

If a vessel is overfilled with cryogenic liquid, then the additional boil-off vapour escaping through the vent-line will create a back-pressure, compressing the liquid. If the additional boil-off is now relieved by opening a second vent or pressure relief valve, the pressure falls, the liquid expands from the already full vessel into the vent-line, and unexpectedly increases the boil-off mass flow causing the vent-line pressure to rise still further. What action should be taken? The first thing to realise is that the problems of overfilling a vessel do not stop when the liquid transfer is cut off. The compressibility of the liquid must be anticipated by slowly opening additional vapour vents, without encouraging liquid to expand into the vent-line.

Secondly, if liquid spills, or more probably sprays, out of the vents, beware of (a) cold burns to personnel, (b) asphyxiation by displacement of breathing air by dense cold vapour, (c) cold damage to carbon steel structures and wet concrete paving.

Therefore it is important to ensure that emergency procedures are in place that permit the immediate and controlled evacuation of personnel to avoid (a) and (b), and the setting up of water sprays to avoid (c).

Clearly, upper liquid level alarms should be set to allow sufficient time for corrective action before overfilling occurs.

### 8.4.2 Stratification: Creation and Detection

Removing liquid at uniform temperature and composition from a vessel or tank is easy. Adding fresh liquid to a tank, whose contents are at the same temperature and composition is also straightforward.

However, these operations are not always so simple in practice. As discussed extensively in Chap. 5 on mixtures, adding fresh liquid to an existing heel is likely to lead to immediate stratification because, invariably, fresh liquid will have a different density.

On the one hand, the fresh liquid could be colder and more dense, in which case it will tend to collect at the bottom of the tank below the heel. On the other hand, the fresh liquid could be less dense (for example, LNG with a higher methane content), in which case it will tend to collect above the heel.

To prevent stratification, the fresh liquid must be adequately mixed with the heel. If it is denser than the heel, then the upper spray inlets should be used to promote mixing. If it is less dense, then the bottom inlets should be used, hopefully with the nozzles pointing upwards and inwards from the tank wall so as to promote stirring motion and consequent mixing.

Once stratification occurs, then problems will inevitably follow.

Pumping out the bottom layer via a bottom exit, or sump exit below the base of the tank, will be without problems until the top layer reaches the exit, when a drop in pressure head and rise in temperature will occur. If the operating margin (see Chap. 6) is too small to accommodate these changes, the inlet pressure will fall below the NPSH. The pump will stall and be unable to empty the tank of the less dense and warmer upper layer.

If nothing is done to remove the stratification, then a rollover with spontaneous mixing will inevitably occur, as outlined in Chap. 5. The rollover itself is not a hazard, but the large increase in boil-off vapour flow certainly is. If the vapour compressors cannot cope, the vapour will have to be released into the atmosphere presenting a potential fire hazard.

Fortunately, in the case of LNG, the vapour plume becomes less dense than air as it warms up above about 150 K, and rises, with rapid mixing, away from any ignition sources.

For higher hydrocarbons, their vapour plumes remain more dense than air as they warm up, and the vapour spreads out at ground level, without significant mixing and dilution with air, presenting a major fire hazard.

### 8.4.3 Stratification: Safe Removal

The problem is how to remove the stratification as simply as possible. It will help, of course, if the tank is instrumented sufficiently well so that the operator knows what he is doing.

The first step is to try and mix the upper and lower layers within the same tank by inducing a vertical stirring action by pumping liquid from the bottom and back into the top through the upper spray inlets. Additional vent gas will be generated, but the rate will be related to the liquid transfer rate. In the absence of other instrumentation, the stratification will have been dissipated when the additional vent gas, produced by heat of mixing, falls to zero.

A more successful technique is to use two storage tanks and circulate liquid between the bottom of one and the top inlet of the second tank. Again, the rate of generation of additional vent gas can be controlled by the liquid transfer rate, and when the additional gas rate falls to zero, the stratification has been dissipated. In principle, there is then no need to continue liquid circulation between tanks, but in practice, operators may be happier to circulate continuously and accept the additional vent gas generated by the heat input of the pumping process.

## 8.5 Safety Laboratory Features

The safety laboratory and associated protocols, at the University of Southampton, were designed to manage and mitigate both the risks posed by low temperature cryogens and also the explosive risks of the fuel gases.

This section will comment on what was done at Southampton with reference to any practice which has evolved in the years since.

Any explosive environment must be zoned in accordance with ATEX directives. Any equipment used in each zone must be specified for use in that zone. ATEX rated equipment is designated a category, each category is corresponds to a zone e.g.: Category 1 is for zone 0, category 2 for zone 1.

In the UK Dangerous Substances and Explosive Atmospheres Regulations (DSEAR) must be adhered to, DSEAR is the UK implementation of ATEX [4].

### *8.5.1 Fire and Explosion Containment*

The structure of the building must be resistant to fire and meet local standards; appropriate explosive environment classification will also be laid out in local legislation (e.g. ATEX in Europe, HazLoc in North America). To minimise the damage to the structure of the building from any explosion venting panels must be included. Venting panels are designed to be the weakest point within the laboratory structure. When subjected to the rapid rise in pressure present during an explosion these panels will open/fail sacrificially to allow the pressure to be released rather than damage the building structure.

To minimise the severity of any fire/explosion the total amount of flammable liquid within the laboratory must be limited.

### 8.5.2 Ventilation

Any laboratory working with cryogenic and/or cylinder gas must be well ventilated. It is recommended to have active extraction. Extraction ducting should be designed so that gas does not accumulate, in neither high nor low points, within the system.

High and low extraction points should be installed; hydrogen and methane will rise, LPG and higher hydrocarbons will fall. It is also worth noting that gases which naturally rise when cold can fall, before rising as they reach ambient temperature.

It is good practice for fans to be linked to gas detectors, when the gas detector is triggered the fan is automatically switched to high flow operation and the laboratory atmosphere is recycled in a matter of seconds. The gas detection system can also be triggered by oxygen monitoring, and the detection of a low oxygen atmosphere will have the same action on the extraction system. Gas detectors should be placed at high, low and bench height—for reasons described in the previous paragraph. Where multiple gases are used, oxygen depletion may be more appropriate than detectors triggered by a single gas.

Any ventilation system, in addition to managing extraction, must also consider vents for incoming gas to replace gas extracted from the laboratory. This may come preferably from fresh air, or from other internal rooms, corridors etc., that do not have risk of gas release.

### 8.5.3 Management of Flammable Gases

A separate and/or dedicated area for use with flammable gas is advised. Laboratory benches with fume hoods with extraction isolated from other fluids, particularly liquid oxygen boil off is preferred.

Any extraction from flammables should be ATEX rated, with electrical equipment (in particular the extractor fan) zoned and coded appropriately.

These arrangements will enable liquid hydrogen, liquid methane, LNG and liquified ethane, propane and butane to be handled in both open topped and closed Dewars, cryostats and containers just as LIN and LOX are.

This will allow researchers to get hands on experience with all cryogenic liquid and experience at first hand some of their peculiar properties, such as those of LNG with it's foaming, spitting and Marangoni wall films.

### 8.5.4 Personnel Safety

Limiting the number of people within the laboratory at any time will reduce the risk of any accident. It also ensures that evacuation through any emergency exit can be achieved promptly.

If static discharge is a potential hazard, i.e. where explosive gases are to be used, antistatic flooring and earthing strips worn on shoes will reduce the risk of electrostatic discharge. An appropriate procedure must ensure visitors wear the antistatic strips as well as staff/students.

Anti-static clothing should be worn, clothing compliant with EN 1149-5 should be sought to fully protect staff.

### *8.5.5 Asphyxiation and Toxic Gases*

An assessment of the asphyxiation risk must be undertaken. The calculation uses the worst case scenario—if all available cryogen was to spill concurrently and generate gas depleting the oxygen level. If the resultant oxygen level is below 19%, then fixed oxygen depletion alarms should be fitted. It is important to remove the volume of any equipment when calculating the room volume as this can be significant.

Portable, battery operated, oxygen monitors are also recommended—these are personal monitors—to be worn; clipped to a lab coat, belt or worn round the neck of the research worker. The oxygen monitors have audible alarms set to signal if the oxygen level drops to 19%. These can be easily tested quite simply by blowing into the detector port—exhaled breath has an oxygen level below 19% and is guaranteed to sound the alarm.

The ventilation procedure outlined in Sect. 8.5.2 should reduce the risk the oxygen levels falling below 20% in the laboratory, to minimise the risk of asphyxiation. Hydrocarbon gases are very slightly toxic and only then at high levels of contamination; the occurrence of which should be prevented by ventilation procedures. Some students have reported headaches after working with propane and butane for periods of many hours.

## 8.6 Particular Single Component Cryogens

The following are personal comments on each of the cryogens, based on the author's continuous working experience since 1954. They should be regarded as additional to the standard advice in safety literature.

### *8.6.1 Helium*

Liquid helium storage Dewars and vessels are fairly robust, by the very nature of their internal insulation and vapour- or liquid-cooled shields. If a vessel full of liquid helium gets knocked over, or perhaps falls off the back of a truck, a lot of cold vapour will be generated. However, there is no need to panic, because it will not blow up. Just put it back into an upright position and save the rest of the liquid helium!

Blockage of transfer lines and necks of storage Dewars from the ingress of air and its solidification below about 50 K is a common experience. Again, there is no need to panic.

Remember that both liquid helium and supercritical helium have high coefficients of compressibility so that a line or neck blockage will **not** lead to a rapid increase in pressure. Indeed, sealed off vessels of LHe can be, and are, sent on long journeys lasting many days without loss of helium.

To get rid of a solid air blockage, it is best to use a soft copper tube, or bunch of copper wires, to introduce some heat and melt the solid air. There is lots of time, so be gentle with your copper probe, and do not break through the neck-wall or line into the vacuum space; and watch out for the sudden plume of cold helium vapour when the blockage melts.

If the blockage is high up in the neck, it could be due to ice, i.e. frozen water. This will require a great deal more heat to melt than a solid air blockage.

A growing problem is the limited resource availability. The easy availability from the US gas fields over the past 50 years has encouraged the wasteful use of applications such as helium-arc welding, party balloons and release at source. The US Bureau of Land Management has recently cut back on exporting helium, to meet the approaching exhaustion of the US gas fields. This has led to helium shortages outside the US, and a growing need for introducing priorities to meet essential uses, including cryogenics.

Recycling helium for cryogenic use has been expanded to reducing rising costs. Also, the growing use of cryocoolers has enabled "dry" cryogenic systems to become standard. For example, the first generation of MRI systems used liquid helium as a coolant. The second generation use 4 K cryocoolers as recondensors to recycle a small amount of helium as a liquid exchange medium.

Art Francis, consultant to the US Bureau of Land Management, was the first to pinpoint the growing shortage in the USA. He also suggested the setting up of a World Bank for Helium as diminishing Earth resource with built-in priorities for access and usage. Today, the helium shortage problem is constantly changing, because of the discovery of new sources in Tanzania and Europe, and the longer development associated with gas fields in East Siberia. (See Ref. [5], pages 305–306 and 313; also Ref. [6]).

Meanwhile, uses such as the helium-arc welding and party balloon industry should have low priority and be stopped from wasting a rare Earth resource.

### *8.6.2 Hydrogen*

Hydrogen can form a combustible/explosive mixture with air between 4 and 75% by volume hydrogen concentration, with an ignition energy one tenth that for methane or natural gas. Any spillage of LH2 or escape of the gas is therefore a potential fire/explosion hazard.

It is therefore a challenge to reconcile these facts with the environmental demands for hydrogen to be used as a motor-car fuel, either as a fuel-cell component or as a replacement for petrol, because its only combustion product is water.

The industrial use of hydrogen under controlled safety conditions is acceptable, but the dispensing of the gas or liquid on the forecourt of a roadside fuel station requires many more built-in safety conditions. Thus, Linde are developing a fully automatic LH2 filling system with no liquid spillage or vapour generation. The safety of LH2 storage in road vehicles is being improved by the use of automatic internal liquid shut-off valves and the development of extremely low boil-off vehicle tanks, so that the need for venting is minimised to once a week or longer.

Ambient temperature vapour spills will generally rise and mix with air to below the 4% flammability limit. However, spills of cold vapour and LH2 will lead to cold plumes of flammable mixture which do not rise and mix. Any ignition source, including electrostatic sparks may initiate a flashback and explosion. The subsequent BLEVE, Boiling Liquid Expanding Vapour Explosion, and fireball would be generated within seconds.

While bulk quantities of 100,000 $m^3$ and more LNG are safely shipped and stored around the world, the necessary insulation for the safe shipping of LH2 requires high vacuum insulation containers.

In fact, the cryogenic technology is not available for the safe handling and shipping of larger volumes in the public domain. Industrial proposals for developing bulk volume shipping and storage of liquid hydrogen should therefore be terminated [7].

### 8.6.3 Neon

With a critical point at 44.4 K and 26.5 bar, neon is an attractive refrigerant under pressure between 27 and 44 K for cooling systems for HTS superconductor applications. While neon is expensive, it has a high latent heat of vaporisation per unit volume (3 times that for LH2). The maximum amount of liquid to be handled is therefore likely to be relatively small, only a few litres.

Neon is, however, a difficult liquid to handle because the temperature and pressure ranges between NBP at 27.05 K and triple point at 24.55 K and 0.43 bar are so small. Thus handling and transferring the liquid can easily result in solid neon being formed.

My experience with solid neon is that once formed, it is difficult to re-liquefy because (a) it has an extremely low thermal conductivity, and (b) heat transfer from liquid to solid neon is sluggish anyway because of the small $\Delta T$.

It follows that an accidental blockage of solid neon in a transfer line is very-difficult to remove without warming up and re-gasifying the whole neon system.

Data on the solubility of impurities in liquid neon is non-existent, as far as I know. The behaviour of He–Ne and $H_2$–Ne mixtures needs to be understood in the separation of high purity neon, but there are no solid mixture phases around the triple point of neon.

The density of liquid neon is 1206 kg/m$^3$ and is therefore 10 and 17 times more dense than LHe and LH2 respectively. Great care is therefore needed if liquid neon is stored in standard LHe or LH2 vessels. With their light-weight neck structures, the vessels may not be strong enough to hold liquid neon and could be easily damaged just by tilting them from the vertical.

If the neck fractures, you will have a problem. The vacuum will be lost, atmospheric air will rush in and condense, and the liquid neon will evaporate rapidly and be lost.

### *8.6.4 Nitrogen*

LIN is the standard, low cost cryogen and is used by everyone. Over-familiarity is the main safety problem, leading to carelessness about the danger of asphyxiation.

By way of example, it should be remembered that a standard 25 l storage vessel of LIN will create 170 m$^3$ of nitrogen gas if the vessel's contents are spilt on to the floor. This volume is comparable to that of a medium sized laboratory or workshop with a floor area of 50 m$^2$ or 550 ft$^2$, giving an idea of how serious a relatively small spillage of LIN can be in a closed space. Additionally, the cold nitrogen gas will displace atmospheric air upwards, rendering the floor and lower levels dangerously deficient in oxygen.

One major property is that LIN boils at a temperature below the condensation temperature of atmospheric air. Thus, any contact with atmospheric air will result in contamination of the LIN by it taking up the gaseous components, together with contaminants like carbon dioxide and water. The boiling temperature of the LIN will therefore rise steadily with time so that the boiling point of the pure liquid at 77.2 K cannot be used as a reliable reference point.

### *8.6.5 Argon*

There is one important point to realise when handling large quantities of LA namely that the asphyxiation hazard of argon is much more serious than that of nitrogen.

The density of argon vapour at ambient temperature is about twice that of air, so a liquid spillage or gas escape will produce a heavy cloud of argon rich vapour in the lower level of an unventilated room. Because of its high density, the vapour cloud will not dilute spontaneously by mixing with air.

Systems designed for LIN and LA are broadly compatible, however attention should be paid to substituting argon for nitrogen, due to the density. Any engineering calculations on pipework, supports, Dewar necks etc. must be double checked due to the mass increase of the working fluid.

### 8.6.6 Oxygen

As a powerful oxidant, LOX presents problems in many areas. Many substances, which are barely combustible in air, will bum fiercely in an oxygen-rich atmosphere.

Hydrocarbons of all types can even be spontaneously explosive. Thus contamination by hydrocarbons of parts, valves, pipework, joints, etc. of a GOX/LOX handling system must be eliminated. Also, vacuum pumps and compressors with oil-lubrication must never be used in contact with oxygen.

LOX spills are a problem because GOX is denser than air. Even a liquid spill in the open air can be hazardous because GOX will flow downhill from the spillage. The GOX will follow the topography of the ground, producing an oxygen-rich atmosphere hundreds of metres away and yielding a totally unexpected occurrence of fierce, oxygen-rich combustion.

Personal clothing, exposed to an oxygen-rich atmosphere, remains highly contaminated with oxygen for a surprisingly long time after returning to atmospheric air. During this period of time, you are a mobile hazard, and lighting a cigarette may result in turning yourself into a human torch!

### 8.6.7 Methane 112.2 K, Ethane 184.2 K, Ethylene 169.2 K, Propane 231 K, n-Butane 272.6 K

As liquids boiling below ambient, all these hydrocarbons have liquid handling properties like those of LIN; with the flammability of their vapours as an additional property. In a fire situation, it is the vapour which burns and so the size of the flame above an open burning pool of liquid is directly related to the evaporation rate. A short time after a spillage on to the ground, the heat flux through the frozen sheath of earth falls significantly and it is only the surface heat flows from the atmosphere and the radiation from the flame which contribute to the liquid evaporation.

Thus, covering the surface of the liquid pool with a foam blanket, to separate liquid and vapour, provides insulation from these two heat flows and an immediate reduction in boil-off rate, thereby smothering the flame.

## 8.7 Particular Cryogenic Liquid Mixtures with Total Mutual Solubility

### *8.7.1 Liquid Air*

The total condensation of atmospheric air leads to a liquid mixture boiling at 78.9 K (see also Sect. 5.1). Subsequent surface evaporation of the liquid air mixture yields a vapour containing less than 7% oxygen, which is insufficient to support life. Thus, liquid air is an asphyxiation hazard from its surface evaporated vapour (NB: not its total boil-off).

The low oxygen content of the vapour results in the oxygen content of the liquid enriching above 21%. As surface evaporation proceeds, the oxygen content of the vapour will rise above 7% so that by the time one half of the liquid has evaporated (by surface evaporation) the oxygen content of vapour and liquid may have risen to 20 and 50% respectively. The vapour may now be safe to breathe without anoxia, while the liquid is a serious oxidant and fire hazard.

The oxygen enrichment, with time, of the liquid can take place in a number of ways. (1) Loss of nitrogen by surface evaporation or boiling of the liquid, (2) Partial condensation of gaseous air on cold surfaces below 81 K, (3) Partial condensation of atmospheric air directly into the liquid.

The oxygen enrichment processes will also be accompanied by auto stratification of a more dense oxygen rich layer (density $\sim$950 kg/m$^3$) below a less dense layer close to 875 kg/m$^3$ of Liquid Air.

The oxygen enriched layer will support combustion much more readily than liquid air. For example, at 25% oxygen, there is a significant increase in burning rate for clothing, textiles, wood, hair etc. At 30% many materials described as fireproof in air may burn fiercely with white hot heat, including leather, nylon, PVC, PUF and fire retardant foams. The energy of ignition decreases strongly with increasing oxygen content so that hydrocarbons, lubricants and solvents may explode almost spontaneously.

Another associated hazard is the fact that the clothing of personnel working in an oxygen rich atmosphere may adsorb the oxygen and become extremely flammable. After leaving the workspace, it is advisable to avoid contact with any ignition source for at least 15 min, like lighting a cigarette. Otherwise, your clothes and skin may catch fire, yielding to a very painful death.

Safety regulations for liquid air recommend that any oxygen enriched liquid above 22% should be treated like LOX, for which there are well established safe handling procedures. This action should STOP the use of Liquid Air being stored for any period from being used other than LOX.

Furthermore, the risk of lower half of a tank being oxygen enriched is too great for any use in any public domain. The use of liquid nitrogen as a cold store would avoid the problems of oxygen enrichment all together.

It is therefore strongly recommended that liquid nitrogen should be used for cryogenic energy storage (CES), NOT liquid air.

It should be noted that the British Cryogenic Safety Manual includes this recommendation in the 2018 5th edition.

Liquid air is coming back as a cheaper cryogen than LIN, but it should be remembered that both vapour and liquid compositions change with time. The only safe way it should be used as a refrigerant is by total evaporation.

### 8.7.2 Oxygen, Nitrogen and Argon

There is a wealth of experience and data on these mixtures and their distillation properties.

In general, all three single component liquids have finite concentrations of the other two in solution. This may not matter in many applications, but there are some where minute concentrations of oxygen or nitrogen in argon may have serious consequences, for example, in the large LA bubble chambers in the LHC at CERN.

### 8.7.3 Liquid Natural Gases LNG

LNG is handled on a very large scale as a standard cryogenic liquid. However, in the laboratory, it exhibits some peculiar properties which make it rather unpleasant to handle, in comparison with, say, LIN or LOX.

Firstly, as a mixture, it exhibits the Marangoni effect, whereby the walls of the containing vessel above the liquid surface are covered in wet fluid and droplets of ethane-rich liquid.

Secondly, it spits and splashes. Looking into an open, insulated bucket of LNG is hazardous because any motion of the low viscosity liquid, produced by touching or moving the bucket, will result in spitting, i.e. a series of small explosions and the ejection of many droplets of liquid into your face, or up to the ceiling if they miss you. This spitting appears to be a property of LNG and not of LCH4 and is presumably due to irregular local homogeneous nucleate boiling or QHN boiling in the LNG meniscus in contact with the wetted walls.

Spillage of LNG on to water, is also accompanied by irregular explosive boiling, again presumably arising from local homogeneous nucleate boiling or QHN boiling. This phenomenon is alarming, but is probably not dangerous in the open air [8]. The evaporated vapour is generally non-polluting because it has a density less than that of atmospheric air above about 170 K (−150 °C), becoming about one half the density of air at ambient temperature. The vapour therefore mixes spontaneously with air to dilute itself below 5%, the lower limit of flammability, and rises rapidly away from possible ignition sources.

On the other hand, spillage into underground tunnels and tunnel networks, like sewage systems, is extremely hazardous and must never be allowed to happen. The vapour is unlikely to be diluted with air to below 5% in the tunnel and a flammable or explosive mixture may be formed. It only requires an ignition source, such as an electric spark, to initiate an underground explosion and widespread damage above ground.

### *8.7.4 Liquefied Petroleum Gases (LPG)*

Liquefied Petroleum Gases are largely mixtures of propane with normal- and iso-butane. Enormous quantities are produced by petroleum refineries as non-condensable gases, which used to be regarded as waste gases and were burnt in flare stacks. But not today because they are in fact a valuable source of energy and now form part of industrial cryogenic activity. The storage, handling and distribution of LPG in very large quantities, as refrigerated liquids down to −42 °C or as ambient temperature liquids under pressures up to 30 bar, is probably more widespread than the whole of the LNG industry.

They do however present a safety hazard, because the vapours are more dense than air by a factor of about two. This means that any spillage, of liquid or vapour, is accompanied by the production of a low-lying cloud of heavy flammable vapour close to the ground or water surface. This cloud does not spontaneously mix with air and dilute itself below the lower limit of flammability and can flow over the ground, guided by topographical features for a long distance. Most ignition sources are at ground level, so the situation is easily reached for a major fire to occur from a relatively small spillage.

If a fire does commence, then the possible disastrous behaviour of a single skinned storage tank of LPG, whether refrigerated or pressurised, must be anticipated and evacuation of the immediate area to a distance of several hundred meters is essential. This is because several LPG tanks have been known to disintegrate in a fire and create a Boiling Liquid Expanding Vapour Explosion, or BLEVE, with very many casualties.

Because of the scale of the LPG industry, very large tanks of 100 m diameter are in common use for storing refrigerated liquid down to −42 °C. The tanks are constructed of pre-stressed concrete, including the roofs, which are designed to withstand storage pressures only slightly above 1 bar, measured in inches water gauge.

Now LPG is a mixture, just like LNG, and suffers from all the evaporation instability and path-dependent mixing behaviours met with LNG as described and discussed at length in Chap. 5. Thus, provision must be made for stratification and its removal, and/or the occurrence of rollover and vapour explosions.

There must be emergency venting to release associated boil-off spikes and thereby avoid over-stressing the concrete structure of the tank, which may not be designed to withstand even the relatively small pressure excursions associated with boil-off instabilities.

## 8.8 Liquid Mixtures with Limited Solubility of Second Component

### *8.8.1 Impurities in Liquid Helium, Hydrogen and Neon*

Generally, solubilities of nitrogen, air, hydrocarbons and water in liquid helium, hydrogen, and neon are expected to be so low, at levels lower than 1 ppb, as to be negligible. Free radicles may be captured by freezing them at temperatures down to liquid helium temperatures, but whether they dissolve may not have been determined by quantitative FTIR work.

### *8.8.2 Impurities in Liquid Nitrogen and Argon*

There are many impurities, including carbon dioxide, with limited solubilities in the 1–100 ppm range at the NBP (see Fig. 5.13a, b and c). Water solubility in LIN was once measured to be 10 ppm at 77 K [9], but subsequent work suggested a lower figure, less than 0.1 ppm, because the FTIR lines for water overlap with a carbon dioxide harmonic [10].

Certainly, LIN tanks and vessels appear to have a lining of white deposit on their wetted walls under many operational conditions.

One safety concern is associated with "pure" argon, which invariably contains low levels of nitrogen or oxygen impurity. In high-energy particle physics, liquid argon is being used in large quantities in bubble chambers where it is exposed to high radiation fields for long periods. Because of the finite levels of nitrogen and oxygen in the liquid argon, it has been predicted that nitrogen diazide $(N_3)_2$ and ozone $O_3$ will form as impurities. Provided such impurities stay in solution in the argon, there is no problem. Out of solution, the solid forms of nitrogen diazide and ozone are known to be pyrophoric, i.e. spontaneously combustible, if not explosive.

Clearly, solubility limits and rates of production of these impurities need to be determined.

### *8.8.3 LOX and Acetylene*

Like liquid nitrogen and argon, liquid oxygen dissolves many impurities in the 1–100 ppm. range at its normal boiling point. But since LOX is such a powerful oxidising agent, the build-up of solid impurity beyond the solubility limit can have serious consequences in the case of anything combustible. One important example is that of acetylene $C_2H_2$, which has a solubility limit generally accepted to be about 5 ppm at 93 K (1.3 bar) [11]. Acetylene can be an impurity in atmospheric air, certainly downwind of industrial complexes, and needs to be removed from the vast quantities of air entering an air separation unit. If acetylene does penetrate the molecular sieve and cryogenic clean-up systems, then it tends to collect in solution in the LOX.

If LOX is removed as a product, then the dissolved acetylene is carried out of the plant. However, if the LOX is not removed, and gaseous oxygen is taken out instead, then the reboiling action of the reboiler/condenser tends to concentrate the acetylene within the LOX bath in the low-pressure column. In which case, the solution concentration rises with time. If it exceeds the solubility limit, then acetylene passes out of solution and collects as a solid in the bottom of the LOX bath. This build-up presents a serious hazard since solid acetylene is pyrophoric, certainly in LOX, and is likely to cause an explosion in the reboiler/condenser.

Steps need to be taken to monitor the presence and concentration of acetylene in the LOX bath by extracting liquid samples for analysis at frequent intervals. In addition, the LOX should be circulated continuously through a secondary hydrocarbon adsorber as a back-up.

### 8.8.4 LOX and Other Hydrocarbons

While the reported solubilities of methane, ethane, and propane in LOX at 90 K are respectively 100, 12, and 1%, higher hydrocarbons have lower solubilities in LOX and present the same problem in principle as acetylene. For a review, see references [12] and [13].

Higher hydrocarbons have solubility limits of 1–100 ppm, and if they are present as gaseous impurities in process air entering an air separation plant, then steps are needed to prevent their build-up as solids in the LOX after penetrating through the clean-up stages.

While the solubilities of methane and ethane in LOX [12] are so high that the solids formation problem may not apply, the presence of percentage quantities of hydrocarbon in a LOX solution is surely a hazard like that of smoke particulates in LOX described in Sect. 8.8.5 below.

### 8.8.5 LOX and Particulates

A major problem arises if hydrocarbon particulates penetrate the clean-up system and collect in the LOX for the following reasons:

(a) Since the hydrocarbon monitoring relies on liquid sampling, it may not detect any build-up of particulate.
(b) The back-up recirculation through the secondary adsorber may, on the one hand, concentrate the particulate at the entrance to the adsorber or, on the other hand, do nothing to remove particulate concentrating in the LOX reboiler.

Monitoring for the presence and build-up of particulate in the LOX reboiler requires in situ cold instrumentation, possibly employing optical fibres, but definitely no electronics, in the LOX.

A fire and explosion in an air separation plant in Malaysia remained a mystery for many weeks after the event, until evidence was found in the wreckage of particulate deposit. The deposit appeared to have arisen from the smoke of forest fires.

## 8.9 Safety of Cryogen-Free Systems

Following the continued development of closed-cycle cryocoolers, with cooling capacities of, for example, 1 W at 4 K, 20 W at 20 K, 100 W at 50 K, 200–1000 W at 90 K, there is a growing trend to cool relatively small cryogenic systems with cryocoolers, in the total absence of any liquid cryogens, whether LHe, LH2, or LIN.

The cold heads are usually placed in a high vacuum, and all heat is transferred by thermal contact and metallic conduction between the superconducting magnet, or instrument to be cooled, and the cold head. Cooldown takes a long time, but the cryocoolers of today are computer controlled and are therefore very convenient to use, without the complication of any cryogenic utilities. What, then, are the disadvantages and safety problems?

First of all, their very convenience will lead to their use by operators with no background in the operation of cryogenic systems. Therefore, some basic training in cryogenics and associated safety procedures needs to be provided beforehand.

Secondly, there is a serious increase in safety hazards associated with cryocoolers which operate below 81 K, the condensation temperature of air. Any air leaking into the vacuum space will condense on surfaces below 81 K and freeze to a solid on surfaces below 50 K. Furthermore, with a cryocooler operating below 50 K, the cryopumping of the air leak will be very good. The vacuum will be maintained and no leak will be suspected, although the cooling performance may be observed to deteriorate.

The hazard now lies with the cryocooler stopping for any reason, such as a power failure, or being deliberately stopped and allowed to warm up. There is little heat capacity for absorbing heat flow through the insulation, and the cryocooler and system will warm up quite rapidly. The cryopumping effect is rapidly lost and any solid air in the vacuum space will turn to vapour. The rate of heat transfer through the insulation will rise and the pressure in the vacuum space may rise very rapidly to one bar, or much higher.

It is therefore important that the vacuum jacket has a pressure relief valve or bursting disc fitted as a standard feature, just like all vacuum-insulated cryogenic liquid containers.

It follows that, before turning off a cryocooler, the vacuum space should be connected to a continuously pumping, high vacuum system, certainly until the temperature of the whole system has warmed up above 81 K.

## 8.10 Summary of Safety Points Raised

1. Read the BCC *Cryogenics Safety Manual: A Guide to Good Practice* and other cryogenic safety literature for a more comprehensive guide to safety.
2. Be aware of the deadly danger of asphyxia, especially when the gases are non-toxic and give no physiological warning of oxygen deficiency, like nitrogen, argon, and hydrocarbons.
3. Do not overfill vessels.
4. Beware of stratification, and the consequences.
5. The increasing use of "liquid air" presents hazards of its own, like asphyxia from breathing the low oxygen boil-off vapour, and oxygen enrichment of the residual liquid.
6. LPG spills are particularly hazardous because they produce heavy flammable vapours which do not readily disperse into the air.
7. Impurities in boiling LIN, LOX, and LA will remain in the residual liquid.
8. High levels of irradiation into so-called high purity cryogens like LA, may generate ozone and diazides from ppb. levels of oxygen and nitrogen impurities.
9. Cryogen-free systems require on-line vacuum pumps when the cryocoolers are turned off to warm-up the systems.

## References

1. The Cryogenics Safety Manual: A Guide to Good Practice, 5th edn. British Cryogenics Council (2018)
2. Zabetakis, M.G.: Safety with Cryogenic Fluids. Plenum Press (1967)
3. Edeskuty, F.J., Stewart, W.: Safety in the Handling of Cryogenic Fluids. Plenum Press (1996)
4. The Dangerous Substances and Explosive Atmospheres Regulations, HMSO (2002)
5. Nuttall, W.J., Clarke, R., Glowacki, B.: The Future of Helium as a Natural Source. Routledge (2012)
6. Cryogenic system for Epe Cavern in Germany. In: Proceedings of the ICEC/ICMC Conference, Oxford (2018)
7. Scurlock, R.G.: Problems of bulk storage and shipping of liquid hydrogen in volumes of 10,000 to 100,000 cubic meters a one bar pressure. In: Proceedings of the IIR Cryogenic Conference, Dresden (2017)
8. Lom, W.L.: Liquefied Natural Gas. Applied Science Publishers, London (1974)
9. Rebiai, R., Rest, A.J., Scurlock, R.G.: The unexpectedly high solubility of water in cryogenic liquids. Nature **305**, 412 (1983)
10. Rest, A.J., Scurlock, R.G., Wu, M.F.: The solubilities of nitrous oxide, $CO_2$, aliphatic ethers and alcohols, and water in cryogenic liquids. Chemical Engineering Journal **43**, 25 (1990)

11. Castle, W.C.: Private communication (2005)
12. Wu, M.F.: The solubility of solutes in cryogenic liquids. Ph.D. thesis, Southampton University (1986)
13. Yun, S.: Experimental and thermodynamical studies on the phase equilibria for $CO_2$ in LNG components at 77–219 K. Ph.D. thesis, Southampton University (1988)

# Index





T. D. Bostock and R. G. Scurlock, *Low-Loss Storage and Handling of Cryogenic Liquids*, International Cryogenics Monograph Series,
https://doi.org/10.1007/978-3-030-10641-6

**D**



**E**



**F**



**G**



**H**



**I**



**J**



**K**



**L**

The manufacturer's authorised representative in the EU is Springer Nature Customer Service Centre GmbH, Europaplatz 3, 69115 Heidelberg, Germany. If you have any concerns regarding our products, please contact ProductSafety@springernature.com

Printed and bound by CPI Group (UK) Ltd, Croydon, CR0 4YY
30/09/2026
02225444-0003